Solar Energy, Mini-Grids and Sustainable Electricity Access

T0239850

This book presents new research on solar mini-grids and the ways they can be designed and implemented to provide equitable and affordable electricity access, while ensuring economic sustainability and replication.

Drawing on a detailed analysis of solar mini-grid projects in Senegal, the book provides invaluable insights into energy provision and accessibility which are highly relevant to Sub-Saharan Africa, and the Global South more generally. Importantly, the book situates mini-grids in rural villages within the context of the broader dynamics of national- and international-level factors, including emerging system innovation and socio-technical transitions to green technologies. The book illustrates typical challenges and potential solutions for practitioners, policymakers, donors, investors and international agencies. It demonstrates the decisive roles of suitable policies and regulations for private-sector-led mini-grids and explains why these policies and regulations must be different from those that are designed as part of an established, centralized electricity regime.

Written by both academics and technology practitioners, this book will be of great interest to those researching and working on energy policy, energy provision and access, solar power and renewable energy, and sustainable development more generally.

Kirsten Ulsrud is a postdoc research fellow in human geography at the Department of Sociology and Human Geography at the University of Oslo, Norway.

Charles Muchunku is an independent renewable energy consultant in Kenya with 15 years of experience in the renewable energy sector in Eastern and Southern Africa.

Debajit Palit is an associate director and senior fellow at the Rural Energy and Livelihoods Division at TERI in India, with 20 years of experience working in the field of clean energy access, rural electrification policy and regulation, distributed generation, and solar photovoltaics.

Gathu Kirubi is a lecturer at the Department of Environmental Sciences at Kenyatta University in Nairobi, Kenya. He holds a PhD from University of California, Berkeley on off-grid rural electrification in Africa.

Routledge Focus on Environment and Sustainability

For more information about this series, please visit: www.routledge.com/
Routledge-Focus-on-Environment-and-Sustainability/book-series/RFES

Solar Energy, Mini-Grids and Sustainable Electricity Access

Practical Experiences, Lessons and Solutions From Senegal

**Kirsten Ulsrud,
Charles Muchunku,
Debajit Palit and Gathu Kirubi**

Routledge
Taylor & Francis Group

LONDON AND NEW YORK

First published 2019
by Routledge

2 Park Square, Milton Park, Abingdon, Oxfordshire OX14 4RN
52 Vanderbilt Avenue, New York, NY 10017

Routledge is an imprint of the Taylor & Francis Group, an informa business

First issued in paperback 2020

British Library Cataloguing-in-Publication Data
A catalogue record for this book is available from the British Library

Library of Congress Cataloging-in-Publication Data
Names: Ulsrud, Kirsten, author. | Muchunku, Charles, author. | Palit,
    Debajit, author. | Kirubi, Gathu, author.
Title: Solar energy, mini-grids and sustainable electricity access :
    practical experiences, lessons and solutions from Senegal /
    Kirsten Ulsrud, Charles Muchunku, Debajit Palit and
    Gathu Kirubi.
Other titles: Routledge focus on environment and sustainability.
Description: New York : Routledge, 2019. | Series: Routledge focus
    on environment and sustainability | Includes bibliographical
    references and index.
Identifiers: LCCN 2018034629 | ISBN 9781138359031 (hardback) |
    ISBN 9780429433955 (ebook) | ISBN 9780429783524
    (mobipocket)
Subjects: LCSH: Solar energy—Senegal. | Microgrids (Smart power
    grids)—Senegal. | Rural electrification—Senegal. | Renewable
    energy sources—Senegal.
Classification: LCC TJ809.97.S38 U47 2019 |
    DDC 621.3124409663—dc23
LC record available at https://lccn.loc.gov/2018034629

ISBN: 978-1-138-35903-1 (hbk)
ISBN: 978-0-367-60670-1 (pbk)

Typeset in Times
by Apex CoVantage, LLC

Contents

Figures

Tables

1 Solar energy, mini-grids, and private sector initiatives

Three students from Germany with an innovative business model for decentralized electricity supply started their own company and went to Senegal with a vision to provide electricity to people in areas without any electricity supply. They created a joint venture with a Senegalese company, hired staff in Senegal, and started to implement solar-energy-based mini-grids in rural villages outside the main electricity grid. Their initiative was different from the already existing small-scale, renewable mini-grids in Senegal, because it was led by a private sector company that invested their own money and took loans to make it possible. They intended to plan, finance, implement, operate, and gradually increase the number of villages they would serve. They expected that they would thereby avoid problems that had occurred in other mini-grids, where those who were responsible for operating them did not have enough incentive to keep them operating when major needs for maintenance would occur. This book analyzes the practical outcomes of this activity and presents lessons that can be built on by others who engage in provision of sustainable electricity access, either practitioners, policymakers, financers, or researchers. The book demonstrates how some people make tremendous efforts to make the world both greener and more equitable through social and technological innovation in practice. Such change agents, through their struggles to change established structures in society, generate knowledge and experience that might provide valuable lessons for other engaged actors and for society as a whole. Belonging in the domain of sustainable energy access, this book analyzes such an example. The example offers a range of lessons on one of the main organizational models for decentralized electricity provision: small-scale mini-grids for rural villages, based on solar energy or other renewable energy sources.

Currently, small-scale renewable mini-grids are among both the most interesting and the most challenging of the decentralized electricity access models. There are intensive innovation struggles in this field, including the efforts of the actors involved in the mini-grid activity analyzed here. There are also high estimates for the number of people who will be best served by this kind of model in the future. In Sub-Saharan Africa, among the 220 million people who are expected to need decentralized solutions, mini-grids are anticipated to cover about two-thirds, according to a scenario developed by IEA (2014, p. 496). They estimate that 315 million people in rural areas of this region will gain access to electricity by 2040, with around 80 million being served by individual systems and around 140 million by mini-grids. This requires the development of between 100,000 and 200,000 mini-grids, depending on the number of households connected to each system.

The main aim of this book is to offer new research on how the implementation, operation, and replication of small-scale solar and hybrid mini-grids can lead to affordable, useful, and sustainable electricity access for large numbers of people. Through its analysis of the practical, real-life experiences involved with this particular type of decentralized electricity supply, this book informs scholarship and innovation in the field and contributes to the larger efforts that are seeking to secure universal access to electricity.

We take the reader through a challenging journey of social and technical innovation and exemplify the real experiences of the committed practitioner. The book illustrates the contrasts between the international celebrations of private-sector-led provision of electricity access and the comprehensive struggles involved in scaling up activities in poor, remote areas and dealing with slow-changing energy sectors. It combines a focus on the electricity systems, business models, and policies with vivid pictures of how the electricity systems are perceived, used, and influenced by village citizens. The book also demonstrates a framework of analysis that can be built on by other researchers who would like to achieve a comprehensive understanding of similar kinds of cases – both energy systems and other kinds of infrastructures, especially at the community or village level.

The book thereby contributes to answering one of the most important "how" questions of our time: How can everyone, across the globe, get access to electricity that is delivered in useful, sustainable, reliable, and affordable ways? The importance of this question hardly has to be explained. It seems self-evident that everyone should have the same right to electricity access. This is also firmly stated in Sustainable Development Goal Seven on universal access to sustainable energy and highlighted by the United Nations' initiative Sustainable Energy for All (SE4ALL). The critical question is: How can this become possible?

Entrepreneurs and driving forces

Three young people formed the company INENSUS GmbH (hereafter Inensus) in 2005, during their university studies in power engineering and power systems in Germany, at the Institute of Electrical Power Engineering at Clausthal University of Technology. The young entrepreneurs had an interest in the African continent and a wish to contribute to that continent in a positive way. They had already achieved familiarity with several African countries through living and working there. The Inensus employee who led the work in Senegal, for instance, had already spent 2 years in Africa before his university studies. The three young men had observed the shortcomings and failures of development aid and were convinced that business-based activities would be much more useful.

A central vision of Inensus was to contribute to electricity access in Africa using a business approach. The initial business idea was to reuse first-generation wind turbines. These were taken down in Europe but were good enough to be reused because 10–15 years of extra use were possible. However, the Inensus founders soon found this to be irrelevant. It would be better to use smaller wind turbines, solar photovoltaics (PV), and diesel generators to provide access to electricity in rural areas. Moreover, after solar PV prices fell, wind was no longer cost effective.

Inensus gradually developed its business to become a provider of total solutions for mini-grids based on innovative technical equipment that it developed. While working on wind, Inensus employees discovered that they had to develop their own devices, for instance, inverters to integrate wind into the system. They also developed a wind and solar monitoring system, which helped in evaluating wind and solar resources and their correlation with one another. During this work, they realized that the interlinkage between the mini-grid customer and the power station were missing because the existing devices (meters) did not have sufficient functions. Thus, they developed a special electricity meter, which became a key device in the mini-grid model implemented through the joint venture they created in Senegal.

When Inensus moved from technical solutions to the implementation of mini-grids in practice, it accomplished this through the development of a business model for mini-grids that they called "The Micro-Power Economy." This was the business model they implemented in Senegal from 2009 onward. On their website, Inensus presented the model as a "business and risk management model for electricity supply to rural villages in developing countries using mainly renewable sources and being based on private investments."

At the time of our finalization of this book, Inensus had grown to include 11 people. Technical development was still an important part of their work, and they continued to have close ties to the university where the three founders had studied. Moreover, consulting on mini-grid initiatives for policymakers, operators, donors, and banks had become an important part of their business. Over more than a decade, Inensus has accomplished the delivery and installation of hardware, as well as consultancy work, in a number of African countries. The company has also started working in Asia. However, as our informant in Inensus told us, their objective was not to be consultants, but to set up and operate mini-grids. They found that it was difficult to survive on this kind of work, but continued to work on this challenge. This book analyzes the results of their attempt to break new ground through their mini-grid activity in Senegal and also shows some of their recent steps to achieve their vision.

Opportunities and challenges for mini-grids

The current number of people without access to electricity is approximately 1.06 billion people according to IEA (2017), most of them living in Africa, Asia, and Latin America. This number has been reduced from 1.7 billion people in 2000 (OECD/IEA 2017). This increased access to electricity is primarily due to centralized grid connections, but over the last 5 years, mini-grids and individual off-grid solutions have become increasingly important, providing 6% of the new electricity connections worldwide between 2012 and 2016 (REN21 2018). Most of the new connections have been made in certain geographical areas such as India, Indonesia, and Bangladesh in Asia and Ethiopia, Ghana, Kenya, and Senegal in Sub-Saharan Africa (IEA 2017). In Africa, about 10% of those with access to some kind of electricity supply get it from decentralized solar photovoltaic (PV) technology (World Bank 2018, p. 30).

Solar mini-grids and other decentralized electricity models offer a potential answer to the question on how to reach all, in addition to conventional grid extension. This is because they have advantages that meet some of the shortcomings of the main electrical grids for providing electricity access in rural areas of low- and medium-income countries (Bazilian et al. 2011; World Bank and IEA 2013; Practical Action 2014; Alstone et al. 2015). While grid infrastructure is expanding, it is difficult to extend the grid to all non-electrified areas, both economically and technically. Conventional grid extension is therefore expected to be feasible for about 40% of the people who are lacking access to electricity, while the remaining 60% will need decentralized solutions, according to the International Energy Agency and the World Bank (2018), both stand-alone systems for individual users and village-level systems for collective use, such as mini-grids.

Another limitation of conventional grid extension is that even if a certain place is connected to the grid and counted as electrified, people face multiple barriers to obtaining and retaining an electrical connection. A common problem is the limited geographical outreach of the grid. The electricity lines often reach just the most central parts of the settlements, and many people thereby live outside the reach of the grid. Moreover, grid lines may pass through the neighborhoods, but if there is no transformer in an area due to the cost considerations of the government, the buildings along the line cannot be connected. Other barriers include an inability to afford a connection or pay the bill for electricity, as well as an unreliable electricity supply (IEA 2011; World Bank and IEA 2013; Winther et al. 2018). The lack of generation capacity and the poor quality of transmission and distribution networks are also barriers, leaving "electrified" populations without a reliable power supply despite being connected to the main grid.

Electricity provision by the use of mini-grids, if they are well designed, implemented, and operated, can fill a gap between conventional grid extension and stand-alone solar systems by compensating for some of the shortcomings of both. They can provide basic, affordable electricity services for people who cannot benefit from stand-alone solar PV systems, and they can provide a high-power electricity supply to areas where conventional grid extension is difficult. Mini-grids consist of a power plant and a distribution grid that operates in isolation from the main electricity grid (IRENA 2017, p. 89). They provide varying levels of electricity services depending on their capacity and technical design, and the electricity generation capacity ranges from around 1 kilowatt (kW) up to 10 megawatt (MW). Some authors divide this into micro-grids (1–10 kW) and mini-grids (the rest, 10 kW–100 MW). However, the size of the mini-grids studied here are on both sides of this split, which is common for mini-grids installed in rural villages. Mini-grids supply electricity to customers from a combination of sources, with or without storage.

Solar energy is emerging as an important technology for mini-grids. In general, solar energy via solar PV technology is the most promising source of electricity for decentralized solutions globally because of its huge resource potential and highly distributed availability and because involved actors do not need to purchase and transport fuel to the installations. Much of the support for activities with solar PV and other renewable energy is also motivated by concerns regarding climate change (Alstone et al. 2015). Although the off-grid use of solar PV technology has started to grow fast in many parts of the world, it still only reaches a limited portion of the people without conventional electricity access (IEA 2014; GOGLA 2015; Bloomberg 2016). Moreover, most of the growth is taking place for very small, individual PV systems, in certain geographical areas, and for certain wealth segments. There is a much larger potential, therefore, for a larger use of solar PV, for a much larger number of users.

In some geographical areas, there is strong competition between solar mini-grids and individual solar systems (solar home systems/stand-alone solar systems and small lighting products such as solar lanterns). It was claimed already in 2009, in a book called *Selling Solar*, that solar mini-grids was a failed model and that the solar technology should be bought and owned by individual consumers (households, enterprises, etc.) "just like generators, motorbikes or washing machines" (Miller 2009). The responsibility for maintenance would also rest with the household or other social unit that owns it. In general, companies selling such individual systems market their products heavily in order to convince people that it is better to own your own system ("finish paying and then have free power thereafter"). Companies or organizations offering electricity services through mini-grids or energy centers try to explain the advantages of larger flexibility of usage, freedom from maintenance, battery replacement, and the purchase of new systems more often than people might expect. The latter is due to breakdowns or lack of replacement batteries (Muchunku et al. 2018). The sales of individual solar systems are leading the competition so far, but the work on delivering electricity services from solar mini-grids is also making significant progress. It is an open question which models will be most important in the future. Certainly, both these kinds of off-grid electricity models, as well as energy centers and charging stations, have advantages and disadvantages and complement each other, and they fit for different purposes and for different geographical contexts.

The progress for solar mini-grids is not so much in making large volumes, so far, but in making significantly better delivery models. However, several challenges remain. For instance, there is a dilemma between affordability of electricity services for the population and economic performance, like for most other electricity models (Ulsrud et al. 2011, 2018; Bhattacharyya and Palit 2016). When people get access to electricity for the first time, their chances to utilize it for a range of different purposes is limited due to economic constraints, affordability of appliances, and limited electricity supply from many of the mini-grids, which is in turn caused by economic and technical constraints that the mini-grid implementers have to handle. There is large potential for further learning and innovation regarding organizational factors, equity considerations, power relations, economic performance, practical factors, policies, and regulations, as further explained below.

A mini-grid case to learn from

This book analyzes a case of private-sector-led, small-scale, renewable-energy-based mini-grids that is relevant for understanding the opportunities and hindrances for such ways of providing access to electricity. It was initiated by Inensus through a joint venture they created in Senegal with the Senegalese partner Matforce CSI (hereafter Matforce). The name of the

Senegal joint venture was ENERSA (hereafter Enersa), which gradually grew from one to five Senegalese employees. The main technology used for electricity generation in the mini-grids was solar PV, while diesel generators were used for backup and additional power generation when needed.

Senegal was selected for this research because it was the setting for the specific mini-grid initiative that we were interested in. The country also has other activities within decentralized use of solar PV, including mini-grids (see Chapter 2). Moreover, like other African countries, Senegal has an abundant solar resource, as well as wind, and thus good preconditions for a transition to a non-fossil energy system.

Our three main reasons for selecting the mini-grid model created by Inensus were first that the model was innovative and advanced and seemed to be a promising example of how this kind of energy model could be designed, operated, maintained, and expanded and how the private sector might contribute to increased electricity access through decentralized electricity systems, given some financial support.

Second, the case strongly illustrates some unresolved issues that hinder private sector contributions to increased electricity access through small-scale, renewable-energy-based mini-grids. The lack of a proper and clear regulatory framework is a common hindrance for private sector mini-grid initiatives, according to practitioners working in different countries in Sub-Saharan Africa, as well as in Asia. In the selected case in Senegal, only six out of 30 planned mini-grids were implemented, because of unfinished and unclear policies and regulations that were promising on paper but not implemented the way they were described. One of these six mini-grids was a pilot project implemented in 2010, and the other five were implemented between July 2014 and March 2015. Creating a well-functioning mini-grid model certainly does not help if politics and regulations do not provide frameworks to enable the large-scale rollout of such models. Problems related to uncertainties in policies and regulations appear to be one of the key obstacles for similar activities in many countries, and the selected case helps in understanding and addressing these.

Third, Inensus, the initiator and main driver of the activity in Senegal, has become one of the most experienced mini-grid experts and project developers working in Sub-Saharan Africa, and they are also active in Asia. We present some of these further activities towards the end of the book. The company has received several awards for their mini-grid business model, including the European Business Award for the Environment in 2012.

For these reasons, we found it instructive and relevant to study how their mini-grids work in practice over time, document Inensus' learning experience, and draw lessons for future work on mini-grids, rural electrification, and efforts towards sustainable energy access for all. Such lessons can be drawn from both the achievements and the challenges they had to face during a highly bureaucratic process in an uncertain policy environment. These lessons are relevant to people who implement and operate mini-grids, policymakers and regulators, donors, researchers, industries that produce technical equipment for mini-grids, and investors in the energy sector. Lessons from one specific context can be used as a basis for further innovation and adaptation to other contexts (Ulsrud et al. 2017). At the same time as we acknowledge that a specific case will always have many special features, we also find that this case is not too special to bring lessons about some basic challenges of mini-grids, not least for Sub-Saharan Africa, as we will further show in our conclusion.

An analytical framework for understanding mini-grids

A mini-grid system has several important dimensions, and these should be studied in combination to get a comprehensive picture of what kind of social unit such an energy system is, which kinds of factors influence how it is implemented, operated, and scaled up, and how it works for the involved actors. In our view, based on our practical experience and previous studies, six main dimensions should be investigated in order to achieve a comprehensive (or holistic) understanding of mini-grids, and in this book, we have devoted one chapter to each of these dimensions. We also show, throughout the book, how they interact in dynamic ways to shape the outcomes of the committed actors' efforts to achieve social improvement. This analytical framework can be relevant also for studies of other decentralized, community- or village-level technology installations and infrastructures.

The dimensions of the analysis are the following six:

1   National and Global Energy System Context
2   Local Context
3   Socio-technical Design
4   Functionality
5   Electricity Services
6   Replicability

Each of these dimensions draws on and combines suitable theoretical approaches to socio-technical change, as we explain below. Such a combination of theories is necessary in order to understand as many factors as possible that affect how the mini-grids have come into being, how they work in practice, and how they might be sustained and replicated. We also build on the previous social science literature on small-scale renewable energy, off-grid rural electrification, and access to electricity for people living in poverty.

The first dimension (National and Global Energy System Context) of the analytical framework zooms out from the mini-grids located in rural villages to the larger energy system context at different scales, including policies and other national and international framework conditions. Such factors are likely to play a role in how local mini-grid projects are designed, operated, and replicated in different countries and regions. Theories on socio-technical systems and transitions to sustainability help in understanding this dimension, including how solar mini-grids can be seen as part of larger processes of emerging transitions to low carbon energy systems (Berkhout et al. 2010). Equity considerations and justice have also been increasingly called for in the literature on such transitions (Jenkins et al. 2018; Leach et al. 2012). These theories highlight the dynamic interaction between technology and society (Ornetzeder and Rohracher 2005; Stirling 2008). Technological advancements influence what a society can do, but a society must also change in order to integrate new ways of using technologies, whether they are electric cars or radically different energy systems. Not only laws and regulations but also economic considerations, ideologies, power relationships, and knowledge systems may have to change in the process. These social dynamics play a role in how energy systems are configured and change. Because such systems are both social and technological, they are called socio-technical systems (Hughes 1983; Bijker and Law 1994).

The conventional socio-technical systems for electricity supply are based on a centralized electricity grid and include established government structures. These energy systems can be seen as socio-technical regimes in the language of transition studies and the multi-level perspective (Geels 2011; Smith and Raven 2012). Such a regime is in a strong position because the technological, institutional, economic, and social elements have developed over long time and become strong and established. They are maintained by strong actors, but they also have weaknesses that create windows of opportunity, such as the inability to reach all. Emerging alternatives, such as decentralized solutions, are usually promoted by other kinds of actors who are in a much weaker position, typical for the socio-technical niches as described as part of the multi-level perspective (Geels 2011; Berkhout et al. 2010; Smith 2007). Long-term efforts by a range of actors are therefore needed in order to build up the alternative socio-technical systems. As pointed out by Ockwell and Byrne (2017), financing hardware/technical equipment is far from sufficient for creating novel energy systems, but such simplified measures are commonly assumed to be the way forward. Three main types of strategies have been shown to be important for such system innovation: learning, building networks, and creating joint expectations (Seyfang and Haxeltine 2012). The learning processes are the most comprehensive among these three (Ulsrud 2015), as this case study serves to illustrate.

The outcomes of such efforts are unpredictable and may even be unsuccessful, and the actors innovate by trying, failing, learning, and trying again. Such learning and building up of novel socio-technical systems takes place both through practical projects, in this literature conceptualized as sustainability experiments or socio-technical experiments, and through attempts to build new institutions or change existing ones. Development of powerful discourses or ways of framing the new technological configurations are also part of strategies used in order to legitimize the novelties and strengthen their opportunities to become normalized and successful (Fuenfschilling and Truffer 2014). In addition to being a case of mini-grids, the Enersa activity can be viewed as a case of this broader kind of phenomenon – pioneering efforts by engaged actors who attempt to create deliberate social and technological change towards a desirable future.

Previous literature on how factors related to this wider energy system context affect small-scale renewable mini-grids have mentioned the role of political ideologies on the role of the state and market, regulations, and various institutions (Bhattacharyya and Palit 2016; Newell and Phillips 2016). A typical feature of the energy sectors in developing countries is that global financing institutions such as the World Bank have pushed strongly for economic liberalization and privatization and created reforms that are still ongoing with unsettled issues and distribution of responsibility between different actors as a result. Other examples are needs for subsidies, vested economic interests and lack of political priority, and limited access to long-term, low-cost capital (IEA 2011; Yadoo and Cruickshank 2012; EUEI PDF 2014; Bhattacharyya and Palit 2016). Such factors can be either part of socio-technical regimes or niches (Geels 2011; Smith and Raven 2012). Both can exist on different geographical scales or levels of governance and work across spatial contexts (Bridge et al. 2013).

The second dimension (Local Context) of the framework concerns the role of the social, spatial, cultural, and material context where people live. This dimension influences how the electricity system works and for whom. This partly depends on how the project implementers adapt the systems to this context and people's practices, needs, and economic situations. Examples of factors that form conditions for electricity provision are settlement patterns, social conditions, and the level of economic activity in the village (Chaurey and Kandpal 2010; Kirubi et al. 2009). Wealth is often very unequally distributed among different social groups in a given place, and this influences affordability and leads to exclusion from access to electricity (Winther 2008; Winther et al. 2018; Leach et al. 2010). It is therefore important to understand the daily struggles of various groups and the hindrances they face for taking advantage of the available electricity services (Ulsrud et al. 2018). The interaction between "the delivery model" of electricity supply and the context in which it becomes introduced and possibly embedded, influences how the systems work in practice and the kind of electricity access the systems give and for whom, when, and where (Ockwell and Byrne 2017; Ulsrud et al. 2011; Rolffs et al. 2015; Winther 2008; Ulsrud 2015; Ahlborg 2017; Rohracher 2003). An important part of the mutual impacts between energy provision and the local context is the interaction between people's practices and the novel socio-technical system, which leads to new and unpredictable practices. What people do with energy is shaped by local knowledge, practical needs, ideas of progress, norms, and values (Shove 2003; Wilhite 2008a, 2008b; Winther 2008; Ulsrud 2015). Emerging practices consisting of changing relations between materials, like technical devices, competences, and meanings, are conducted and reconstructed in everyday life due to the repetitive character of social life, including emerging energy provision (Shove 2003; Winther 2008; Smits 2015). The practices in turn influence the performance of the electricity provision (Ulsrud et al. 2011).

The third dimension (Socio-technical Design) of our framework concerns the specific, intended process of planning, designing, and implementing the energy system, and thereby the details of the socio-technical design (or configuration) of the energy system as intended by the implementing actors. Not only the macro-level energy systems mentioned above, but also the decentralized energy systems studied at the micro-level, such as mini-grids, can be seen as socio-technical systems, and it is easy to see that a mini-grid in a village is a socio-technical configuration, or system, composed by a range of social and technical elements. The social and technical elements of such a system cannot be separated (Williams and Sørensen 2002; Russell and Williams 2002). Among the elements that are more social than technical are the types of energy services provided (e.g., lighting, phone charging, TV, grinding), operational routines, ownership, financing arrangements, tariff setting, payment arrangements for electricity fees, the actors' roles and responsibilities, education, and knowledge required to operate the technology, and the rules for use of electricity. These have organizational, socio-cultural, practical, and political aspects. The motivation and interests of the involved parties, as well as the power relationships between them, are also examples of the social elements of a mini-grid system, as are leadership styles and trust (Ulsrud 2015). The technological elements are also many, although not as many as the social elements. Technical elements include electricity meters (like a special meter that was very central in the mini-grids studied in Senegal), inverters, light bulbs, sockets, and solar panels. In order to understand why the system was designed and implemented in a certain way, the considerations of the involved actors are important. This dimension can be viewed as the planning and implementation of a socio-technical experiment. Their space for maneuvering is influenced by dimensions 1 and 2.

The fourth dimension *(Functionality)* is about the actual functioning of the mini-grid systems. The way in which a socio-technical configuration functions in practice always differs from how it was planned and anticipated (Russell and Williams 2002), and it is shaped by the interaction between technical and social elements of the system, the interaction between the involved actors, as well as interaction and embedding between the system and the contextual dimensions. In other words, the outcomes of sustainability experiments are uncertain and contingent on a range of factors. This dimension is to a large extent about the challenges and struggles of project implementers and helps in understanding the perspective of those actors that struggle to create social and technological change, in this case those who develop, operate, or support mini-grids.

Learning processes after implementation are unpredictable, iterative processes between technical and social elements. For instance, the users of technologies, such as the electricity subscribers, operators of the electricity provision, and the administrators develop their own practices and in this way affect the socio-technical system (Ornetzeder and Rohracher 2005; Williams and Sørensen 2002). Technological change is a social process in which social actors, technology, and institutions interact with and challenge one another. This leads to vigorous learning, innovation, and adaptation but also, and more often than not, it leads to resistance, setbacks, breakdowns, and disappointments. Such processes are influenced by the broader social context, such as the history and culture of specific geographical areas (Späth and Rohracher 2012), as mentioned under dimension 2 and the larger energy systems as mentioned under dimension 1. Societal trends not related to electricity supply are likely to play a role, including historical developments in a region or country (socio-technical landscapes) (Geels 2011). This dimension especially concerns the operational and economic sustainability of the mini-grid systems, which are common goals and challenges for energy provision (Alzola et al. 2009; Camblong et al. 2009; Ulsrud et al. 2011; Bellanca et al. 2013), including mini-grids. Operational sustainability can be defined as the system's ability to have continuous operation and maintenance, while economic sustainability can be defined as the system's ability to cover the costs of operation and maintenance and create a surplus for expansion (Ulsrud et al. 2018).

The fifth dimension (Electricity Services) of the framework is about the access to and qualities of electricity services for different groups, created by the help of the mini-grids. This dimension concerns the final outcome of a local mini-grid system – who receives access, where, when, how, and why. The UN has developed a tracking framework, called the Multitier framework, for measuring the progress of global energy access (ESMAP 2015). The framework classifies different levels of electricity access based on the power and hours of electricity provided, which tend to differ between off-grid solutions and the grid. The qualities of the electricity services are also crucial, including physical accessibility in different spatial areas, affordability, practical usability, and reliability. Not only practical aspects of the electricity use but also the arrangements and schedules for making payments for electricity play a role for user satisfaction, as this case study shows. It is crucial to view such factors from the users' and nonusers' own perspectives. This dimension is influenced by all the other dimensions mentioned above.

The sixth and final dimension (Replicability) concerns the factors that influence the possibility for scaling up or replicating the system. In discussions about mini-grids and other off-grid electricity systems, the concepts of "replication" and "upscaling" are used to refer to ways of moving on from a pilot project or a small number of projects to a larger number of projects or business units (Bhattacharyya 2014). These concepts are further applied to achieving widespread and common use of an energy model, such as widespread use of mini-grids based on renewable energy technologies. In contrast, within theories on transitions to sustainable socio-technical systems, as mentioned in the introduction, the concept of "upscaling" is sometimes used synonymously with "transition," which is a much larger social and technological change. A transition is far more than increasing an innovative energy model from a few to 30, 50, or 100 units, although such an increase would likely be a large effort and entail years of struggles to overcome many hurdles. A transition, as explained in the literature on transitions toward sustainability, is a longer-term process wherein the dominant ways to produce and use energy (or to provide other important social functions) become radically different (Coenen et al. 2010). Replication and upscaling of off-grid energy models are, potentially, steps on the way toward transitions, but this cannot be known for sure because innovation processes are not linear and cannot be predicted (Russell and Williams 2002).

An example of large-scale, long-term transitions would be if off-grid renewable energy solutions came to be a mainstream, normalized way of providing electricity to significant parts of the population, and that a range of institutional arrangements, actors, and technological solutions were in place, including new government offices, laws, regulations, and school curricula as well as people's preferences, ideals, knowledge, and views on the normal ways of doing things. The development of such novel socio-technical systems seems to have started in the field of electricity provision. Pilot projects such as the one Inensus has done via Enersa in Senegal are the most important drivers for such transitions, and this is not true only if they achieve their own goals. According to the literature, they normally encounter some problems. However, a central function of such experiments, even though the project implementers have to abandon certain visions, is that they help society to learn and explore potential solutions for the future (Brown et al. 2003; Raven 2005). For instance, these experiments demonstrate both partial solutions and unresolved issues, including the choices that policymakers and regulators have to make to achieve more progress on rural electrification and sustainable energy for all. Even a pilot project that never becomes more than a pilot contributes to such socio-technical learning processes – socio-technical system innovation. Nevertheless, a small warning is required against the idea that it does not matter if a project stops working. It truly does matter for those in a rural community who participate in a pilot, benefit from it, and might not be able to keep it operating completely on their own over a long time or to restart it after a breakdown (see Ahlborg 2015).

In this analysis, we focus on the replication of projects, and we acknowledge that this may not contribute to large-scale, comprehensive transitions to sustainable energy systems. Replication is used herein to refer to shifting from a few examples of an innovative energy model to a larger number, as Enersa hoped to do in Senegal by expanding from one mini-grid to 30 and potentially continuing to increase the numbers later. Replication in "large numbers" for micro- or mini-grids is here taken to mean 25–30 systems or more for one project implementer, because this means a significant amount of investment, effort, and operational follow-up.

"Replication" is not a perfect word, because it implies a direct copying of an existing model, but we use it for lack of something better and because the alternative "upscaling" less emphasizes learning from one project to another. The literature on social and technological innovation makes it clear that there is hardly ever a direct mechanical copying of a socio-technical configuration (as a certain mini-grid model). There will always be learning from experience, and innovation will always take place during the process of building on lessons from other projects and the effort to move an activity forward.

Large-scale replication of village-level projects can be difficult. It is suggested by Palit (2013) that implementation of off-grid projects in clusters can assist in the management of the projects. An example is found in Chhattisgarh state in India, where Chhattisgarh Renewable Energy

Development Agency (CREDA) runs a large number (several hundreds) of small solar plants, including small mini-grids. The operation and maintenance is organized according to clusters of 10–15 villages, supervised by a mobile technician who assists the operators in each power plant (Millinger et al. 2012). Parallel challenges can be found in community energy projects in Europe. Projects may be difficult to replicate in a large number because they are designed to be small-scale and rooted geographically. Such projects are radically different from mainstream solutions. Therefore, it might be difficult to achieve wide diffusion without reformulating and reinterpreting them, through standardization and simplification (Seyfang and Smith 2007; Seyfang et al. 2013).

We bring the dimensions together in the following *research question*: Which factors influence the achievements of small-scale renewable private-sector-led mini-grids, and how? Three kinds of desirable achievements for mini-grid systems are considered:

- They function well in practice, in terms of long-term operational and economic sustainability (relates to dimension 4)
- They provide good quality electricity access (affordable, accessible, reliable, and useful) (relates to dimension 5)
- They can be replicated in large numbers (relates to dimension 6)

Research approach – emphasizing qualitative methods

Social science based and inter-disciplinary research is suitable in order to deepen the understanding of how new kinds of energy models can be developed and embedded in the social fabric. It is also evident from this and other case studies concerning how different energy models work that the details matter. We have therefore performed a detailed analysis, providing a rich picture of the experiences of the people involved and what happened in the encounter between rural villages in Senegal and an innovative and committed mini-grid implementer. With this book, we describe our case study as it happened in order to show the diversity and richness of the case itself. We compare the case with earlier studies where relevant and combine a deep understanding of the particular case with drawing conclusions that are likely to have broader relevance, both for other mini-grid initiatives and for the provision of electricity access in general.

Comprehensive case studies are important for the understanding of how society can move toward new kinds of energy systems in the future and capture different kinds of dimensions as the ones described above. Case studies on the ground are necessary because a rich understanding of individual projects and factors at multiple societal scales that influence them can contribute in important ways to debates about larger changes in energy systems and how transitions to sustainability can take place.

We collected most of the data for this case study in March 2016 in the Thies region in Senegal, in four of the six villages to which Enersa supplied electricity through mini-grids. Our research team consisted of social scientists and technical experts from Africa, Asia, and Europe. We interviewed 65 men and women in households and 25 key people in the villages like mini-grid operators, electricity committee members, and village leaders. Key informant interviews were also conducted in the cities of Thies and Dakar with nine government officials in four government units and a donor, and later by email, phone, and meetings during the writing of this book. We did about 45 hours of interviews with Inensus. The selected villages were Sine Moussa Abdou, Ndombil, Léona, and Maka Sarr.

We emphasized qualitative methods in order to understand different actors' perspectives and motivations, the hindrances they face, and the factors that influence their struggle to create novel social structures. The interview questions were adjusted as the fieldwork evolved over time and we gained better understanding of what was important to investigate in order to answer the research questions. Some interviews were carried out in random groups into which people had gathered in public places or compounds. In addition to this qualitative research, we carried out a quantitative survey to quantify some of the observations, especially in order to know how widespread certain views and experiences were among the citizens, document some socioeconomic characteristics such as education levels, and get an overview of choices made by various groups with regards to the use of energy.

Rather than evaluating the mini-grid project and the type of electricity access it provides based on certain indicators of its performance and achievements, we present an in-depth understanding of what has happened and why, as seen from the perspectives of the various people involved. In addition, it is still useful to have an element of evaluation or assessment. This is because it makes sense to discuss what has functioned effectively or not in order to understand why.

The structure of the book

Our six-step analytical framework structures our analysis and shows the journey of the key actor, Inensus. Chapter 2 concerns Inensus' struggles to overcome barriers of unfinished regulations and national and global politics of electricity supply. Chapter 3 describes the local communities where Inensus would attempt to meet people's needs. Chapter 4 concerns how Inensus designed the social and technological features of their energy system design (the business model) and how they took the local context into account, as well as the strategy they used for the implementation and operation. Chapter 5 shows how the mini-grids actually worked in practice in terms of operations, maintenance, and other aspects that influenced the long-term sustainability of the power supply. Chapter 6 presents the users' perspectives on how the services worked, discusses the factors that influenced their experiences, and shows who obtained access and why. Importantly, Chapter 7 discusses whether and how it would be possible to replicate similar solar and hybrid mini-grids in larger numbers of villages, either by Inensus themselves and their partners or by other actors. Chapter 8 presents lessons likely to be relevant for other initiatives on solar and hybrid mini-grids, both in Sub-Saharan Africa and elsewhere. Chapter 9 concludes with the insights this case study brings into the conditions for implementing and sustaining this kind of electricity model, emphasizing those factors or social structures that are mainly outside the control of the key actors. The chapter also presents recommendations for policymakers, donors, and financing institutions and expresses concerns regarding the approach of upcoming large mini-grid projects in Senegal and elsewhere.

References

Ahlborg, H. (2015). Walking along the lines of power: A systems approach to understanding co-emergence of society, technology and nature in processes of rural electrification. *PhD thesis*. Göteborg, Sweden: Chalmers University of Technology, Department of Energy and Environment.

Ahlborg, H. (2017). Towards a conceptualization of power in energy transitions. *Environmental Innovation and Societal Transitions*. http://dx.doi.org/10.1016.2017.01.004

Alstone, P., Gershenson, D. & Kammen, D. M. (2015). Decentralized energy systems for clean electricity access. *Nature Climate Change*, 5: 305–314.

Alzola, J. A., Vechiu, I., Camblong, H., Santos, M., Sall, M. & Sow, G. (2009). Microgrids project, Part 2: Design of an electrification kit with high content of renewable energy sources in Senegal. *Renewable Energy*, 34 (10): 2151–2159.

Bazilian, M., Nussbaume, P., Rogner, H., Brew-Hammond, A., Foster, V., Pachauri, S., Williams, E., Howells, M., Nyiongabo, P., Musaba, L., Gallachoir, B. Ó., Radka, M. & Kammen, D. (2011). Energy access scenarios to 2030 for the power sector in Sub-Saharan Africa. *Utilities Policy*, 20: 1–16.

Bellanca, R., Bloomfield, E. & Rai, K. (2013). *Delivering energy for development: Models for achieving energy access for the world's poor*. Rugby, UK: Practical Action Publishing.

Berkhout, F., Verbong, G., Wieczorek, A. J., Raven, R., Lebel, L. & Bai, X. (2010). Sustainability experiments in Asia: Innovations shaping alternative development pathways? *Environmental Science and Policy*, 13 (4): 261–271.

Bhattacharyya, S. C. (2014). Business issues for mini-grid-based electrification in developing countries. In Bhattacharyya, S. C. & Palit, D. (eds.) *Mini-grids for rural electrification in developing countries: Analysis and case studies from South Asia*. Switzerland: Springer.

Bhattacharyya, S. C. & Palit, D. (2016). Mini-grid based off-grid electrification to enhance electricity access in developing countries: What policies may be required? *Energy Policy*, 94: 166–178.

Bijker, W. & Law, J. (1994). General introduction. In Bijker, W. & Law, J. (eds.) *Shaping technology, building society: Studies in sociotechnical change*. Cambridge, MA: Massachusetts Institute of Technology Press.

Bloomberg New Energy Finance. (2016). Off-Grid Solar Market Trends Report 2016. Market Analysis.

Bridge, G., Bouzarovski, S., Bradshaw, M. & Eyre, N. (2013). Geographies of energy transition: Space, place and the low-carbon economy. *Energy Policy*, 53: 331–340.

Brown, H. S., Vergragt, P., Green, K. & Berchicci, L. (2003). Learning for sustainability transition through bounded socio-technical experiments in personal mobility. *Technology Analysis & Strategic Management*, 15 (3): 291–315.

Chaurey, A. & Kandpal, T. C. (2010). A techno-economic comparison of rural electrification based on solar home systems and PV microgrids. *Energy Policy*, 38 (6): 3118–3129.

Coenen, L., Raven, R. & Verbong, G. (2010). Local niche experimentation in energy transitions: A theoretical and empirical exploration of proximity advantages and disadvantages. *Technology in Society*, 32 (4): 295–302.

ESMAP. (2015). Beyond connections: Energy access redefined. ESMAP/Sustainable Energy 4 All (SE4ALL). Technical Report 008/15.

EUEI PDF. (2014). *Mini-grid policy toolkit: Policy and business frameworks for successful mini-grid roll-outs*. Eschborn: European Union Energy Initiative Partnership Dialogue Facility (EUEI PDF).

Geels, F. W. (2011). The multi-level perspective on sustainability transitions: Responses to seven criticisms. *Environmental Innovation and Societal Transitions*, 1 (1): 24–40.

GOGLA. (2015). Lighting Global and Berenschot. Global Solar Off-Grid Semi Annual Market Report July–December 2015.

Hughes, T. P. (1983). *Networks of power: Electrification in western society, 1880–1930*. Baltimore: Johns Hopkins University Press. 474 pp.

IEA. (2011). Energy for all: Financing access for the poor. Special excerpt of World Energy Outlook 2011, First presented at the Energy for All conference in Oslo, Norway in October 2011: OECD/IEA.

IEA. (2014). World Energy Outlook: International Energy Agency, Paris. www.iea. org/publications/freepublications/publication/WEO2014.pdf

IEA. (2017). Energy Access Outlook: International Energy Agency, Paris. www.iea. org/media/topics/energyaccess/database/WEO2017Electricitydatabase/

IRENA. (2017). *Rethinking energy 2017: Accelerating the global energy transformation*. Abu Dhabi: International Renewable Energy Agency.

Jenkins, K., Sovacool, B. K. & McCauley, D. (2018). Humanizing sociotechnical transitions through energy justice: An ethical framework for global transformative change. *Energy Policy*, 117: 66–74.

Kirubi, C., Jacobson, A., Kammen, D. M. & Mills, A. (2009). Community-based electric micro-grids can contribute to rural development: Evidence from Kenya. *World Development*, 37 (7): 1208–1221.

Leach, M., Rockström, J., Raskin, P., Scoones, I., Stirling, A. C., Smith, A., Thompson, J., Millstone, E., Ely, A., Arond, E., et al. (2012). Transforming Innovation for Sustainability. *Ecology and Society*, 17 (2).

Leach, M., Scoones, I. & Stirling, A. (2010). *Dynamic sustainabilities: Technology, environment, social justice*. London: Earthscan. 212 pp.

Miller, D. (2009). *Selling solar: The diffusion of renewable energy in emerging markets*. London: Earthscan. XXVII, 306 s. 99 pp.

Millinger, M., Mårlind, T. & Ahlgren, E. O. (2012). Evaluation of Indian rural solar electrification: A case study in Chhattisgarh. *Energy for Sustainable Development*, 16 (4): 486–492.

Muchunku, C., Ulsrud, K., Palit, D. & Jonker-Klunne, W. (2018). Diffusion of solar in East Africa: What can be learned from private sector delivery models? *WIREs Energy Environ.* https://doi.org/10.1002/wene.282

Newell, P. & Phillips, J. (2016). Neoliberal energy transitions in the South: Kenyan experiences. *Geoforum*, 74: 39–48.

Ockwell, D. & Byrne, R. (2017). *Sustainable energy for all: Innovation, technology and pro poor green transformations*. New York: Routledge.

Ornetzeder, M. & Rohracher, H. (2005). Social learning, innovation and sustainable technology. In Filho, W. L. (ed.) *Handbook of sustainability research*, pp. 147–175. Frankfurt: Peter Lang.

Palit, D. (2013). Solar energy programs for rural electrification: Experiences and lessons from South Asia. *Energy for Sustainable Development*, 17 (3): 270–279.

Practical Action. (2014). *Poor people's energy outlook 2014: Key messages on energy for poverty alleviation*. Rugby, UK: Practical Action Publishing.

Raven, R. P. J. M. (2005). *Strategic niche management for biomass: A comparative study on the experimental introduction of bioenergy technologies in the Netherlands and Denmark*. Eindhoven Centre for Innovation Studies, VDM Verlag. 340 pp.

REN21. (2018). Renewables 2018 Global Status Report. REN21 Secretariat, Paris. www.ren21.net/gsr-2018/chapters/chapter_04/

Rohracher, H. (2003). The role of users in the social shaping of environmental technologies. *Innovation: The European Journal of Social Science Research*, 16 (2): 177–192.

Rolffs, P., Ockwell, D. & Byrne, R. (2015). Beyond technology and finance: Pay-as-you-go sustainable energy access and theories of social change. *Environment and Planning A*, 47: 2609–2627.

Russell, S. & Williams, R. (2002). Social shaping of technology: Frameworks, findings and implications for policy with glossary of social shaping concepts. In Williams, R. & Sørensen, K. H. (eds.) *Shaping technology, guiding policy: Concepts, spaces and tools*. Cheltenham: Edward Elgar.

Seyfang, G. & Haxeltine, A. (2012). Growing grassroots innovations: Exploring the role of community-based initiatives in governing sustainable energy transitions. *Environment and Planning C: Government and Policy*, 30 (3): 381–400.

Seyfang, G., Hielscher, S., Hargreaves, T., Martiskainen, M. & Smith, A. (2013). A grassroots sustainable energy niche? Reflections from community energy case studies. *S3 Working Paper 2013-21*. UK: University of East Anglia.

Seyfang, G. & Smith, A. (2007). Grassroots innovations for sustainable development: Towards a new research and policy agenda. *Environmental Politics*, 16 (4): 584–603.

Shove, E. (2003). *Comfort, cleanliness and convenience: The social organization of normality*. Oxford and New York: Berg Publishers.

Smith, A. (2007). Translating sustainabilities between Green Niches and sociotechnical regimes. *Technology Analysis & Strategic Management*, 19 (4): 427–450.

Smith, A. & Raven, R. (2012). What is protective space? Reconsidering niches in transitions to sustainability. *Research Policy*, 41 (6): 1025–1036.

Smits, M. (2015). *Southeast Asian energy transitions: Between modernity and sustainability*. Farnham, UK: Ashgate Publishing.

Späth, P. & Rohracher, H. (2012). Local demonstrations for global transitions: Dynamics across governance levels fostering socio-technical regime change towards sustainability. *European Planning Studies*, 20 (3): 461–479.

Stirling, A. (2008). Science, precaution, and the politics of technological risk: Converging implications in evolutionary and social scientific perspectives. *Annals of the New York Academy of Sciences*, 1128: 95–110.

Ulsrud, K. (2015). Village-level solar power in practice: Transfer of socio-technical innovations between India and Kenya. *PhD dissertation*. Department of Sociology and Human Geography, Faculty of Social Sciences, University of Oslo.

Ulsrud, K., Rohracher, H. & Muchunku, C. (2017). Spatial transfer of innovations: South-South learning on village-scale solar power supply between India and Kenya. *Energy Policy*, 114: 89–97.

Ulsrud, K., Rohracher, H., Winther, T., Muchunku, C. & Palit, D. (2018). Pathways to electricity for all: What makes village-scale solar power successful? *Energy Research and Social Science*, 44: 32–40.

Ulsrud, K., Winther, T., Palit, D., Rohracher, H. & Sandgren, J. (2011). The Solar Transitions research on solar mini-grids in India: Learning from local cases of innovative socio-technical systems. *Energy for Sustainable Development*, 15 (3): 293–303.

Wilhite, H. (2008a). *Consumption and the transformation of everyday life: A view from South India*. London: Palgrave Macmillan.

Wilhite, H. (2008b). New thinking on the agentive relationship between end-use technologies and energy-using practices. *Energy Efficiency*, 1: 121–130.

Williams, R. & Sørensen, K. H. (eds.). (2002). *Shaping technology, guiding policy: Concepts, spaces and tools*. Cheltenham: Edward Elgar. 404 pp.

Winther, T. (2008). *The impact of electricity: Development, desires and dilemmas*. Oxford, UK: Berghahn Books.

Winther, T., Ulsrud, K. & Saini, A. (2018). Solar powered electricity access: Implications for women's empowerment in rural Kenya. *Energy Research and Social Science*, 44: 61–74.

World Bank. (2018). *Tracking SDG7: The energy progress report 2018: A joint report of the agencies: IEA, IRENA, UN Statistics Division, the World Bank Group and the WHO*. Washington, DC: World Bank.

World Bank & IEA. (2013). *Global tracking framework*. Sustainable Energy for All: World Bank and International Energy Agency. 289 pp.

Yadoo, A. & Cruickshank, H. (2012). The role for low carbon electrification technologies in poverty reduction and climate change strategies: A focus on renewable energy mini-grids with case studies in Nepal, Peru and Kenya. *Energy Policy*, 42: 591–602.

2 Just come and invest! The energy system context

Inensus, with its vision to contribute to positive change in Africa through a business approach, chose Senegal as the country for implementation of their mini-grid model for several reasons. The market potential was large, the wind and solar energy resources were good, and there was interest in mini-grids. Another factor that led to their choice was that the German government's development agency Deutsche Gesellschaft für Internationale Zusammenarbeit (GIZ) had a program in the country, which could offer support. The final reason was that the government of Senegal had started to develop a regulatory framework that looked promising and was promoted as investor friendly. From the outset, there were unclear or unfinished regulations for tariff setting, but Inensus anticipated, based on the feedback from their contacts in Senegal, that these issues would be solved progressively in the course of project implementation. "Just come and invest" was the message from the Ministry of Energy and Development of Renewable Energy (MEDER) in Senegal. In theoretical terms, there was an emerging niche for small-scale, renewable-energy-based mini-grids in Senegal, where Inensus could try to put their new technology and innovative business model into practice.

The Republic of Senegal is located in the West African Sahel and is among the 15 member states of the Economic Community of West African States (ECOWAS). The country has an estimated population of 15.3 million (as of 2016–17) and an area of 196,722 km^2.[1] Senegal's macroeconomic performance during 2015 and 2016 was strong, with the economy growing at 6.5% and 6.6%, respectively, making Senegal the second fastest growing economy in West Africa and the fourth fastest in Sub-Saharan Africa as a whole. In terms of real GDP, the country experienced a growth of around 4.5% during 2014 (IMF 2015). However, 46.7% of Senegal's national population and 57.3% of its rural population were living in poverty in 2011 (IMF 2012).

As per the World Energy Outlook 2016 report, a little over 39% of the population in Senegal (approximately 5.78 million people) lacked access to electricity in 2014 (IEA 2016). There is a huge disparity between the urban and rural areas, with 88% of the population in urban areas having electricity versus only 40% in the rural areas (or about 30%, according to interviews with officials in the Ministry of Energy). Senegal has a better electricity access rate relative to many other Sub-Saharan countries, particularly its neighboring countries, where the rural access rate ranges between 2% and 20%. As in many other countries, it is not clear from the statistics whether the access figure includes the population with mini-grid connections or off-grid stand-alone systems. Thus, there may be some ambiguity in the number of people who have electricity. The western portion of Senegal, which includes the regions of Dakar, Thies, and Diourbel and has higher population densities, is better suited for electricity supply by the main electricity grid. The main grid is usually a viable option where population densities are high. However, similar to other countries in Sub-Saharan Africa, Senegal's electricity sector is characterized by growing demand and lack of reliable supply. In the southern regions of Kolda and Tambacounda, diesel generators are common in many urban centers and large rural communities. Some of these places are linked by large diesel-powered mini-grids.

The characteristics of the power sector in Senegal

The conditions for mini-grids in Senegal are influenced by both the general characteristics of the power sector and some particular policies directed toward mini-grid developers. The electricity sector in Senegal is governed by MEDER. This Ministry sets the targets for the sector and implements them through four institutions with the following responsibilities:

1   SENELEC (Société National d'Éléctricité; The National Electricity Utility) generates and distributes electricity. Independent power producers that provide electricity for injection to the grid can only sell to SENELEC.
2   CRSE (Commission de Régulation du Secteur de l'Electricité; Commission for the Regulation of the Electricity Sector) regulates the electricity sector, including the definition of tariffs and awarding of concessions and licenses.
3   ASER (Agence Sénégalaise d'Electrification Rurale; Senegalese Agency for Rural Electrification) is mandated by the Ministry to implement rural electrification in order to increase access to electricity and reduce poverty.

4   ANER (Agence nationale pour l'énergie renouvelable; National Agency for Renewable Energy) has a mandate to promote the use of renewable energy in all activity sectors (in particular, agriculture, health, education, and livestock production).

SENELEC manages the high-voltage transmission network (about 238 km), which delivers electricity to the major distribution centers. The private sector can be involved in distribution of electricity through the medium-voltage (around 7,553 km) and low-voltage (around 6,761 km) networks that supply electricity to the end consumers (Sanoh et al. 2012).

Influenced by the World Bank, Senegal was one of the first countries in Sub-Saharan Africa to introduce private sector participation in their electricity sector. This reorganization was initiated through two electricity laws[2] in 1998. These laws describe the way in which the electricity sector in Senegal would be organized, together with an LPDSE (Lettre de Dèveloppement du Secteur de l'Energie; Energy Sector Development Paper). Before this time, SENELEC held a monopoly over the generation, transmission, and distribution of electricity. The electricity laws removed SENELEC's monopoly and transferred some of the responsibility for rural electrification from SENELEC to ASER. The reform opened up for private sector investment in electricity production, and some companies started setting up generating stations to become independent private power producers. In 2015, of the 886 MW installed capacity, 57.4% was nationally owned and 42.6% belonged to the private power producers (SENELEC 2015). However, lack of investment in the electricity sector is a major challenge, according to our interviews with officials in this sector.

Law 98-29[3] was amended in 2002 to increase the transparency of tender procedures and thereby make them more attractive for private sector actors. LPDSE was changed in 2003 and 2008 to place stronger emphasis on the role of renewable energies.

A more recent LPDSE, which is currently guiding the electricity sector, was adopted by the government in 2012 and covers the period from 2013 to 2017. The paper states the following targets:

a   achieving energy security and energy access for all;
b   combining thermal generation, bio-energy, coal, gas, and renewables in the energy mix and developing regional interconnections;
c   further liberalizing the energy sector through facilitating independent power generation and institutional reform;
d   reducing the cost of energy and lowering subsidies by increasing the competition in the sector; and
e   improving energy sector regulation.

Specific targets for this plan include reducing the electricity production cost to an average of 60–80 XOF/kWh[4] and increasing the portion of renewable energy to 10%. The plan also includes a coal-fired plant (125 MW), several gas power plants, a hydroelectric station, and several solar and wind power plants. Positive views on this were expressed by our informants in the government. Thus, renewable energy has gradually come into the plans. When we asked officials at ASER's renewable energy department how they see the future role of renewable energy technologies in Senegal, they answered that solar is the most important due to the good conditions. They also mentioned the good wind conditions along the coast and that plans have been made to connect some wind power to the grid. In the field of solar PV, some donor-supported activities have been carried out on off-grid electricity provision, including projects on health centers and schools. Communication networks have also been extended by the use of solar PV technology. A medium-size 30 MW solar power plant was under planning at the time of our fieldwork, and the contract had been signed with the company so that installation could start.[5] This would be grid-connected solar PV. The Ministry officials we interviewed also mentioned other renewable energy projects they had been involved in, including solar home systems and solar pumps with German cooperation, a 5 kW solar power station with Spain, and two solar power stations with Japan. They also mentioned South–South cooperation with India on solar home systems. The electricity generation mix in Senegal consists of about 67% from diesel, around 9% hydropower, 9% gas, around 2% renewable energy sources (PV, wind, biomass, solar), and the rest from thermal, co-generation, and imports.[6] There is a law on feed-in tariffs for grid-connected renewable electricity generation, but it has been neither implemented nor developed in detail, and we were informed that this is a long process.

PASER: the Senegalese rural electrification plan of action

Provision of electricity access in rural areas, led by the rural electrification agency, ASER, is guided by a plan called PASER ("Plan d'Action Senegalais d'Electrification Rurale"; Action Plan for Rural Electrification in Senegal). The plan has strengthened the focus on rural electrification as a national priority and decentralized power supplies are part of it, but its implementation has been slow (Mawhood and Gross 2014). The plan consists of the three programs listed below, and the first two are relevant here.

• PPER (Programme Prioritaire de l'Électrification Rurale; Rural Electrification Priority Program);

- ERIL (Électrification Rurale d'Initiative Locale; Local Initiative for Rural Electrification); and
- PREM (Programme Énergétique Multisectoriel; Multi-sectorial Energy Program).

The first of the three, PPER, is Senegal's major program for rural electrification and is implemented by ASER. It is based on an idea of giving concessions for electricity supply to companies that then receive responsibility for electrification and distribution of electricity for certain geographical areas. The approach was designed as part of the mobilization of private sector investors in rural electrification, which was strongly influenced by the World Bank through their financial and technical support (ESMAP 2007). Through this program, Senegal has been divided into ten regional-scale concession areas for provision of electricity services. The program is coordinated through ASER's department for concessions. ASER announces tenders for the concessions, inviting private sector companies to submit bids and compete for the concessions. The company that offers to connect the highest number of customers by the use of a governmental output-based subsidy wins the competition, and the concession contract lasts for a period of 25 years (Mawhood and Gross 2014). If the companies use renewable technologies, an additional subsidy is available, thus supporting a renewable energy niche, which is still small, as shown by the energy mix mentioned above (De Gouvello and Kumar 2007).

Six of the ten concession areas have been allocated. For each concession area, the concessionaire company develops a local electrification plan, including the technology to be used, taking into account the uncertainty of demand and the geographic conditions (Sanoh et al. 2012). Mini-grids powered by solar PV technology can be included, as can other off-grid renewable energy applications, such as solar home systems. There are many villages far from the grid where there are no current plans to offer the main grid, according to our interview with Ministry officials. The concession plan further defines other objectives, such as the type of investment, ways of facilitating productive use of electricity, and connection of public infrastructure. International donors cover up to 80% of the investment cost (through the government), and 20% must be covered by the concessionaire (a private sector operator). Most of the companies that have a concession are foreign, large companies. Some are partly owned by foreign governments, others are multinational companies. According to Ministry representatives, in some areas the company does not invest anything.

The way in which the electricity tariffs are set is a central and challenging issue, which is the responsibility of the national regulator (CRSE), but also influenced by broader political processes in the country. The World

Bank has had a strong impact on the structure for tariff setting for private sector electricity supply through concessions in Senegal, and this structure was described with frustration by our informants in the energy sector.

The regulator calculates the tariff for the concessions, the customer tariff for SENELEC consumers, and the bulk purchase price that the concessionaires must pay to SENELEC when they purchase electricity to be distributed within their area. The tariff structure (used for the concessions) has four levels: three flat rates based on power level (very small consumers, small consumers, medium consumers) and a meter-based tariff for larger consumers. Power level is the same as installed power capacity, measured in watts (W), kilowatts (kW), etc. The monthly customer bill also includes a "payment facility" for spreading out the costs of connection, internal wiring, and energy-efficient lamps. Officials in the energy sector clearly expressed that the tariff structure is not good, and that "it was made by some World Bank consultant" many years back. One of the problems they mentioned was that the tariffs for the small consumers entail a higher price per kWh than those for larger consumers.

A complicating factor cited in our interview with the Ministry was that for political reasons, SENELEC has continued to own existing lines in some concession areas, although the initial plan was that ASER would manage everything for rural electrification from 1998. The companies that hold the concessions own the grid extensions they build, and they meter the power from the point where their line links with the SENELEC line. This has led to parallel supply to different customers and variations in tariffs. A problem pointed out by the Ministry was that the SENELEC price is lower than that of the concessionaires. This seemed to contribute to a wish to change the tariff structure and harmonize tariffs – in other words, get a system where everyone pays the same price per unit of electricity.

The six CERs (Concessions d' Electrification Rurale; Rural Electrification Concessions) that have been allocated so far are the following:

1   St. Louis/Dagana: COMASEL (Compagnie Marocco-Sénégalaise d'Electricité/Louga SAU; COMASEL Louga S.A.). This is a subsidiary of the Office National de l'Electricité (ONE), Morocco's electricity utility company.
2   Podor/Louga/Linguère: COMASEL.
3   Kaffrine/Tambacounda/Kédougou: ERA (Energie Rurale Africaine; ERA SA.) This is a Senegalese joint venture of France's electricity utility EDF and the Senegalese company MATFORCE CSI. (www. erasenegal.com/)
4   Mbour: STL.

5   Kolda/Vélingara: ENCO. This is a multinational firm headquartered in
     Dakar, Senegal. (www.enco-services.com/)
6   Kaolack/Nioro/Fatick/Gossas: ENCO.

The progress of the six concessionaires and thus the results of imposing an
ideology based on liberalization and private sector involvement in order to
solve the problems of electricity sector is mixed. According to ASER staff,
only two (COMASEL and ERA) have managed to achieve significant progress in the number of electricity connections through the national grid and
decentralized solutions, several years after starting the work. Causes for
delays are both the lengthy administrative processes and the tariff model,
which is said to hinder profitable operation. Not only government officials,
but also private sector actors and development partners express doubt about
the current approach to rural electrification, including the tariff structure
mentioned above. A revision of the approach is ongoing, but there is
little progress. The doubt seems to be the reason for the delay in allocating
the four remaining concessions, based on indications during our interview
with MEDER.

   In addition to the concession approach, both ASER and SENELEC continued with the historical model of state-funded electrification (Mawhood
and Gross 2014). SENELEC still plays an important role in the distribution
of electricity both within and outside the concession areas, despite the
reforms, and they still have a type of monopoly, according to interviews.
SENELEC chooses to focus on the urban electricity supply, where the revenues from the customers are best, while ASER is responsible for rural
electrification (Sanoh et al. 2012). However, SENELEC operates in uneco-
nomical ways, as it has done for a long time, according to studies (Sanoh
et al. 2012; Mawhood and Gross 2014). The tariffs are insufficient to cover
the costs, and SENELEC is therefore heavily dependent on subsidies. This
situation prevents SENELEC from investing in maintenance and less fuel-
intensive generating plants, according to the Renewable Energy and
Energy Efficiency Partnership (REEEP) (2014)[7] and Sanoh et al. (2012).
Both SENELEC and independent power producers depend primarily on
imported fuel (Sanoh et al. 2012). The electricity generation cost in
Senegal therefore strongly depends on the volatility in the world market
oil price. In the absence of adequate capital, electricity provision is
expanded primarily in areas already covered by or close to the existing
network. Rural areas in particular are falling behind in electrification
because they are expensive to electrify and people in them usually purchase
little electricity.

   ASER also implements a so-called emergency program.[8] It is designed
similarly to the historical state-funded electrification model (Mawhood and

Gross 2014). The first emergency program was implemented from 2008 to 2012, and the second from 2014 to 2018. Thus the PPER program, which is the dominating part of Senegalese rural electrification strategy, facilitates foreign direct investment (FDI) by large companies and a kind of a government monopoly side by side in Senegal. Donors channel money through the emergency programs for specific projects that they want to promote. In addition, the program described below, ERIL, is the part that looks to small actors. Most of the structures described above can be seen as an established, slowly changing socio-technical regime that primarily maintains the conventional solutions, centralized grid extension based on fossil fuels, but that has started to promote various renewable energy technologies, especially for supply to the main grid. No clear-cut division exists between government units that maintain established regime systems and those that are influencing niches for off-grid solutions, including mini-grids.

The ERIL program – for mini-grids and other decentralized solutions in individual villages

The second PASER program mentioned earlier, ERIL, invites projects and businesses to implement and operate decentralized solutions in individual villages. This call is what the German start-up company Inensus responded to when they started the Enersa mini-grid activity in Senegal. Mini-grids are a possible option for the concessionaires as well, but the mini-grids are then part of a larger electrification plan for a concession area through grid and off-grid solutions. Only large companies such as the ones listed above (utility companies and other significant actors) have so far taken on these tasks. In contrast, the approach of ERIL is to give "mini-concessions" to companies or organizations that take the lead on provision of electricity services in one or several villages through decentralized solutions, such as mini-grids and solar home systems. One could say that the program is more specifically designed for niche technologies such as emerging small-scale, renewable, off-grid ways of providing access to electricity.

The ERIL program was developed because of the lack of progress under the PPER program in order to increase the electrification rate through decentralized solutions, according to a key informant with detailed knowledge about the Senegalese energy sector. Thus, weaknesses of the dominating regime contributed to creating windows of opportunity for alternative socio-technical configurations and system innovation that could in turn contribute to broader access to electricity and better utilization of renewable energies. These would usually be promoted by other actors than mainstream solutions, often weaker and less supported by existing institutional structures. ERIL initiatives may be implemented by private companies, NGOs, community groups, or a consortium thereof. Since ASER's approach to rural electrification is technology neutral, renewable energy can be combined with other fuels, such as diesel, in both the PPER and the ERIL programs. ASER has the responsibility to accompany the operators to apply for an ERIL contract with MEDER, for 15-year licenses for electricity sales and 25-year concessions for electricity distribution.

A rural electrification fund (FER; Fond d'Électrification Rurale) has been established to support the ERIL projects, but the fund's planned levy of 0.7% on all electricity sold nationwide has only been passed in law, not implemented. This means that cross-subsidization from electricity sales to provision of electricity for those people who do not have access is still in its infancy in Senegal. This mechanism is well established in Kenya, for instance, starting in 1978, in terms of cross-subsidization between programs carried out by the government itself. This mechanism in Kenya was developed for conventional, large mini-grids based on diesel generators, but it can possibly be used for small-scale, private-sector-led, renewable-energy-based mini-grids, with some adaptations.

The implementation of policies in Senegal, including the ERIL program, takes place through decrees. The decrees that regulate ERIL are based on the main policy documents described above, the electricity laws, and the LPDSE. The following two decrees concern the development of a mini-grid market in Senegal through ERIL:

• Arrêté 002674, signed on March 14, 2011: Defines the modes of financing of ERIL operators through ASER.
• Arrêté 002675, signed on March 14, 2011: Defines all procedures related to ERIL, including applications from project implementers for licenses and contracts to implement and operate mini-grids or other decentralized electricity provision methods.

These decrees set the procedure for actors who want to apply for an ERIL license. The company or other actor (including NGOs) prepares an application for the ERIL license and concession contract, accompanied by ASER, and submits it to MEDER. It is then transmitted to CRSE which has the power to regulate mini-grids as stated in the decrees. CRSE is then supposed to evaluate the economic feasibility of the project within 45 days, based on the operator's business plan and the concerned community's ability to pay.

Ideally, CRSE sets the individual maximum tariffs for the concerned project. Different tariffs can be issued to different projects, allowing an internal rate of return of 12% on the private investment. After setting the tariffs and preparing the contract, the regulator is to return the relevant documents to MEDER. During another period of a maximum of 45 days, the

Ministry is then to prepare a draft ERIL license and concession contract, which will thereafter be negotiated with the applicant. When both parties have agreed, the contract is to be signed by the Minister of Energy and the applicant's CEO.

In reality, however, this procedure has only been carried out once, and instead of the few months allotted to the process, as explained above, it took 4 years. In fact, the only extant ERIL license was granted to the Enersa pilot project. This was where Inensus' main problems in Senegal occurred and they ended up implementing only six out of 30 planned mini-grids.

Inensus and Enersa's encounter with the ERIL program

Inensus, on behalf of its joint venture with Matforce, Enersa, took the initiative to apply for licenses under the ERIL program at the earliest opportunity. They began to meet with the regulator from the moment that the decrees described above were signed and presented by the government, because this was the time when the ERIL program could be implemented.

However, this became a four-year-long, exhaustive process for Inensus and Enersa. They made a comprehensive and patient effort on the way toward the first and only license, and for long periods, this was a full-time job for one person.

Inensus had frequent and long meetings with the regulator over these 4 years in order to establish a tariff model that could both be approved by the regulator and work for the Enersa mini-grids. Together with the regulator, they gradually worked out a tariff model that could work for them, at the same time as it was relatively close to the model for the concession program created by "the World Bank consultant" many years back. This was done in order to build on a model that was familiar for the regulator.

The tariff model that was approved by CRSE for the concession and license contract between the MEDER and Enersa for the pilot mini-grid was based on limitation of both electrical power (load) and energy consumption (units of electricity or kWh). The license contract subdivides the electricity customers into four different service levels, and in the electricity provision offered in the villages, these levels were used as a basis for letting people choose a level that was suitable and affordable to them, as we explain in Chapter 4.

- Service level 1: Maximum power 50 W and maximum energy consumption of 6 kWh per month
- Service level 2: Maximum power 100 W and maximum energy consumption of 12 kWh per month

- Service level 3: Maximum power 150 W and maximum energy consumption of 18 kWh per month
- Service level 4: Power of more than 150 W and energy consumption metered per kWh

Moreover, the contract allowed for a cost-reflective tariff and set a maximum price per kWh to be charged by the company (669 XOF per kWh, approximately 1.02€).

When they finally received the license for their pilot project in the village Sine Moussa Abdou in 2014, the project implementers saw this as a major breakthrough for their work in Senegal. They were sure that they would now have a good chance to obtain the other licenses they needed for the following mini-grids, including five more that they had already installed. However, the opposite happened. By the end of 2015, government officials indicated that no further licenses would be issued under the conditions described above. This was because the government was planning to have a national uniform electricity tariff in Senegal, and this would have major implications for the mini-grid sector, as will be explained below.

In addition, Inensus had the impression that the regulatory commission had certain reservations toward making decisions about such ERIL license contracts on behalf of MEDER under the prevailing conditions. The complicated and prolonged process to issue the first and to this day only ERIL license in Senegal seemed to scare the regulatory commission off from working out similar contracts for other mini-grids. Inensus' long-term effort to contribute to new regulations suitable for small-scale, private-sector-led, renewable-energy-based mini-grids can be seen as an effort for institutionalization of niche electricity provision, also called institutional work (Fuenfschilling and Truffer 2014). They were quite alone in this work, which made it especially difficult. This is reminiscent of a point by scholars of niche development, who find that one of the important processes that strengthen niches is the formation of networks of actors. Other actors, such as ASER and GIZ, although involved in mini-grid development, did not actively contribute to the license contracts and tariff setting process at this stage, considering this solely the regulator's responsibility.

The electricity tariff setting – a political and complicated issue

The government's argument for creating a uniform tariff for provision of electricity (so-called tariff harmonization) was to ensure equity, so that rural customers supplied by concession holders or ERIL projects would not pay more for electricity than SENELEC customers did. Ministry representatives told us that this is a problem of justice and that people in rural and urban areas should not pay different tariffs.

An explicit policy statement in favor of differentiation of tariffs would be very unpopular among voters. Tariff setting is therefore a highly political issue. As expressed by CRSE representatives, tariffs are political, and the government is therefore promising to reduce tariffs. They also said the government's considerations around tariff setting are not based on a goal to cover the real costs, but rather on a goal to make tariffs affordable, for instance looking at people's capacity to pay based on kerosene consumption. Officials in CRSE explained to us that one of the main challenges in the sector is that many people are not able to pay the electricity tariffs. Another informant, from the donor side, said that in urban areas, SENELEC has to charge tariffs that do not spark extensive protests. He also argued that private solutions in rural areas would not be accepted because people do not want to pay higher tariffs than the general ones and that this is because they have heard promises from the politicians.

Based on the problematic situation explained above, the government has given CRSE an instruction to develop a new tariff model where all electricity customers pay the same tariff. The kind of electricity system the customer is connected to, either grid or off-grid, will not make a difference. The idea is to make the SENELEC tariff (120 XOF/kWh, about 0.18 €) the benchmark tariff. However, this is lower than the cost-reflective tariff needed by private sector mini-grid implementers to make their business viable so that such actors can diffuse their services.

The government has a preliminary plan for continued involvement of private sector operators, including mini-grid operators. The idea is to compensate them for the difference between the harmonized tariff and the tariff they would have to charge to make the operation profitable, to cover the so-called financial viability gap. CRSE officials told us that the government would support the mini-grid operators in obtaining the tariffs they need. However, it is not clear where the money will come from. The fund for rural electrification mentioned above does not yet have money. According to these officials in CRSE, one of their main challenges is that they do not have a good method for determining tariffs. They told us that they are working on a new method to harmonize the tariffs but have not yet determined how to do it.

The Ministry indicated that they were undertaking a study to determine the viability gap for the whole electricity sector and mentioned that an initial estimate was an annual cost of 3.5 billion XOF per year (5.3 million €). The calculation was based on estimated targets of all concessionaires' connections, estimated electricity needs, and the cost of electricity as supplied by the concessionaire. Even with a proposed subsidy to address the viability gap, there were still doubts as to how the subsidy would be correctly determined. Moreover, from the perspective of Inensus, there would be tremendous uncertainty and fear that the government would default or delay in making its subsidy (viability gap) payment to the concessionaires or ERIL operators.

Three paradoxes related to small-scale, private-sector-led, renewable-energy-based mini-grids in Senegal

The ambiguous situation described above is currently slowing down or completely preventing private investment in small-scale, renewable-energy-based mini-grids in Senegal, and there are three related paradoxes. The first is that the utility tariff from SENELEC, to which the cost-reflective tariff of the private sector company is compared, is subsidized. According to our interviews with insiders in the electricity sector, SENELEC has rarely been profitable for many years. We were told that nobody wants SENELEC to fail, so when SENELEC has needed support, the government or donors such as the World Bank has stepped in.

A second paradox is that approximately 150 mini-grids are operating without a license, including the five projects that Enersa implemented as a start of their planned upscaling in Senegal. Some of the other mini-grids have been implemented under a donor program for support to rural electrification, the Program for the Promotion of Renewable Energy, Rural Electrification, and Sustainable Supply of Household Fuels (PERACOD). PERACOD created a project called The Rural Electrification Senegal (ERSEN) jointly by GIZ and ASER. ERSEN supported Senegalese companies for the implementation of hybrid renewable energy mini-grids in remote areas using solar and diesel power (EUEI PDF 2014). A total of 80% of the mini-grid investment cost was financed by GIZ, 10% was financed by customers, and the 10% balance was provided by the private company that would be responsible for the operation. This differed from the Enersa projects, where the private company invested much more, owned and operated the project, and carried the main risk of the activity, as will be explained in Chapter 4.

The ERSEN mini-grids are owned by the government, and the operation and maintenance is assigned to private operators. The private operators are meant to run the mini-grids and sell electricity produced at regulated prices. They are supposed to do this based on an ERIL concession and license contract with the MEDER, but no applications have been submitted to the MEDER under the ERSEN project.

Thus far, the respective authorities have not acted upon the widespread operation of mini-grids without the legally required licenses. This might not be a significant problem for the donor-led projects, but for a private-sector-led activity, such as Inensus' initiative through Enersa in Senegal, it is a major issue. Those investors that committed to the Enersa business, for instance, could not accept being involved in a project operating without legal approval from the government. Therefore, they withdrew immediately when the chance for further licensing was definitively closed.

A third paradox is that at the same time as Enersa's expansion in Senegal has been blocked by regulatory issues, large international donor funds are flowing into Senegal for implementation of hundreds of mini-grids. About 400 mini-grids are under planning and implementation in different regions in Senegal as part of the "emergency program." Some mini-grids are 100% solar powered, and others are solar-diesel hybrids. Public tenders will be used to identify companies that will build the power plants and small grids. These mini-grids are not dependent on private investment since they are mainly subsidized, commonly around 90% and up to 100%. Of these, some are funded by the United Nations Development Programme (UNDP), some by the International Renewable Energy Agency (IRENA), and some by the Islamic Bank, according to interviews with government officials in the energy sector. Such inflow of capital into mini-grid installation is a current international trend. For certain donor-funded projects, the funds will be channeled through the government, and the installation will be achieved through government tenders. Other projects are implemented directly by the respective donors. Private sector companies are contracted to do the implementation, including supply of equipment. The operation is accomplished through separate tenders for operation of the mini-grids, including maintenance, such as battery replacements.

This implementation strategy is similar to that of the previous solar and hybrid mini-grids in Senegal, including the ones supported by GIZ. Many of these have already broken down because they were abandoned by the operators, according to GIZ. The operators are not willing to make large investments in maintenance, especially battery replacement, or install more capacity. With operators abandoning existing donor-funded mini-grids, it is a significant concern that a similar model is being used for the new and large mini-grid initiatives. A Ministry representative mentioned an alternative strategy for operation that could prevent breakdown. He said that they want to train the local population to operate and maintain mini-grids. However, the long-term operation is thereby likely to depend on the follow-up over time by the donors or the government and by economic support to the operators.

Some Ministry officials mentioned problems for mini-grids. "The main problem is to maintain the batteries," they told us. They found that a renewable power supply integrated into the main grid was much easier to maintain, since there is no need for storage of electricity and thus no battery.

Another problem for mini-grids, they said, is that many people do not have an income, due to socioeconomic conditions. They wanted to rebuild the model (probably referring to the tariff model) to improve affordability. One person argued that the grid would be extended everywhere in the end, and that off-grid solutions would therefore not be needed in the future.

Inensus expressed that donors could contribute to the private sector's initiatives in new ways. Instead of mainly focusing on technical assistance and financing, they could also involve themselves in solving problems related to policies and regulations. A hindrance to such involvement might be the mainly technical and economic expertise employed by GIZ and similar donor agencies. In our interview with a person working on the PERACOD program, he touched upon the question of what GIZ could have done to influence the regulatory issues that forced Inensus to walk away. He felt that GIZ was too small a donor to have any impact. He said that those with bigger budgets, such as the World Bank, would be listened to; however, he feels that they are not trying to solve the problems and instead accept the situation and want the government to be dependent on their money because of underlying interests larger than energy. They want to have influence on the country in general. Other large donors, such as UNDP, have a similar approach, we were told. GIZ can, in this situation, be seen as a niche actor that attempts to support emerging socio-technical configurations that can contribute to increased access to electricity in new and sustainable ways at the same time as they support German entrepreneurship and industry abroad. The World Bank is a more mixed actor according to this description, most of all maintaining business as usual, and perhaps not being concerned about the poor performance of the Senegalese electricity sector, because the interests it represents might see it as useful that developing countries are not independent enough to make their own decisions. However, it is not possible in this case to go deeply into these power relations and interests, apart from describing the observations of the actors in the energy sector in Senegal.

Pico-solar systems and solar home systems meet fewer barriers

Because of the complications described above, sale and service of pico-solar systems and solar home systems (SHS) seems to be a more promising field for businesses in Senegal than mini-grids, and they are important in areas with a population too dispersed for mini-grids. Such systems are not hindered by regulations and appear to have less competition from donor-funded projects. The PERACOD program, which is supported by

GIZ, mentioned to us that they wanted to give technical assistance to small businesses and help develop a market for solar products in Senegal, including a good business environment. One concern for businesses is high taxation, which leads to high prices and then to low competitiveness compared with equipment from neighboring countries where taxation is lower. The Ministry of Finance signed a five-year renewable energy convention with ASER, providing tax exemptions for equipment to ASER's rural electrification projects, including renewable energy technologies (Lighting Africa 2012). Full exemption on all solar products, including household systems, is being considered in relation to discussions in the parliament on a new act on rural electrification.

Summary of the energy system context's role

Influenced by the World Bank, the government of Senegal had started to invite private sector involvement in rural electrification policies in Senegal, both for small and large concessions. For Inensus and their Enersa, the policy framework looked good in the beginning. However, they later realized that in practice, the framework became a major stumbling block to their mini-grid initiative and to further private-sector-led initiatives in Senegal.

The ERIL licensing process did not function at all the way the policy documents described it, and the company had to spend considerable time with government officials to identify a solution. When the government thereafter declared that it would not issue any more licenses of this kind, Enersa's investors withdrew, and 24 planned mini-grids had to be canceled.

The chapter has shown important reasons for the complicated and wearisome process that the small company Inensus had to undergo in Senegal and the negative outcome. The direct reason was the government's decision to harmonize tariffs. Two interesting questions are why this decision was made and why cost-reflective tariffs for private-sector-led mini-grids were not seen as an option.

From the company's perspective, the alternative of cost-reflective tariffs was very much justifiable: Private-sector-operated mini-grids would serve people in areas where the main grid was not possible and where mini-grids would be the only opportunity for electricity provision beyond SHS. The tariffs, despite being high enough to cover the cost for providing the electricity (minus support received for the investment costs), would still provide cheaper energy services of much better quality than the solutions used before, such as kerosene, candles, torches, and dry cell batteries.

Moreover, from a private sector company's perspective, it is also self-evident that tariffs (or any other prices for products or services delivered by a business) would have to reflect the costs of operating the business.

In addition, cost-reflective tariffs and private sector investment has been part of the liberalization and privatization doctrine behind the comprehensive energy sector reforms pushed through in developing countries by the World Bank and other international agencies from the 1980s onward. Neo-liberal economic ideas are well-established elements in the current discourse on the importance of private sector involvement in the energy sector. Also Inensus represents such ideas, but based on other visions and values than those that have guided the international agencies.

From the government of Senegal's perspective, the issue of tariffs in off-grid electricity provision appears very different, and several factors might have played a role in the government's conclusion that all tariffs have to be the same per unit of electricity. The views on renewable energy and off-grid electricity supply in general seemed to be relatively positive among the officials we interviewed in MEDER, ASER, and CRSE. As an emerging alternative over the last two to three decades, renewable energy, including off-grid models, was still a relatively new and relatively marginal part of rural electrification (i.e., an emerging niche technology). However, the government's view was that those who would be supplied by decentralized renewables should not pay more per unit of electricity than those who would be connected to the main grid. These officials had also observed problems; for instance, some mini-grids had stopped operating due to battery issues. Some of them gave hints that they saw grid extension and grid-connected renewables as easier than off-grid solutions.

One trigger for the strong drive toward harmonization of tariffs seemed to be the existing tariff structure for concession-based electrification by large, some of them multinational, companies, which had led to frustrations and problems for the government. The Senegalese experience was thereby that differentiated tariffs (as based on a World Bank consultant's suggestions in the 1990s) were not a good system. This negative experience from the concessions system seems to have strengthened the skepticism toward allowing any off-grid project to have cost-reflective tariffs. In turn, this skepticism affected Inensus' committed initiative very negatively.

The chapter has further pointed to the strongly political nature of tariffs, and this can be viewed as an important part of the political economy of electricity provision. Politicians promise lower tariffs to get votes, for instance, as in many other countries. This political tendency likely reflects an important reason for the decision to have the same tariff everywhere in the country.

For an individual and relatively small donor, it seems difficult to influence or assist with policies and regulations because these are influenced by underlying global political and economic interests. Such donors have nevertheless had some impact on the situation in Senegal by the kinds of projects they have promoted and supported. They have also thereby affected the situation that the small company Inensus and their partners found themselves in when they tried to do what they were encouraged to do – create business models that could make it possible to have the private sector involved in providing electricity access to people outside the reach of the main electricity grid. Many high-level forums on off-grid renewable energy and access to electricity have called for such initiatives. Inensus also saw private-sector-led business models as the most viable solution for creating change in poor communities. Inensus was convinced that services delivered by innovative and efficient private companies that invested their own money in the projects would be best because these companies would work hard to create long-lived and dynamic solutions. The large donors, especially the World Bank, started to promote private sector involvement a long time ago and considered laws and policies, but it seems as if they have neither provided enough assistance on finding out how the policies could be implemented in practice nor had an interest in doing so.

Affordable tariffs are important, as was argued by the government officials when they explained the need for uniform tariffs. The task of providing affordable access to electricity for all is a challenge of much larger dimensions than what is generally understood according to our previous research (Ulsrud et al. 2015, 2018). The same is true for the oft-mentioned visions for economic progress, more jobs, and income generation by the help of electricity. However, the questions of uniform tariffs, subsidies, or private sector involvement are only a small part of these larger topics, as will be shown in the following chapters.

Notes

1 www.worldbank.org/en/country/senegal/overview
2 The laws were dated January 29, 1998, and April 14, 1998.
3 The law of January 29, 1998.
4 The Senegalese currency is West African Franc (XOF). 1 Euro (€) approximately equals 656 XOF (March 2016).
5 Interview with staff at ASER's department for innovation and renewable energies.
6 SENELEC Annual Report 2015. www.senelec.sn/images/rapportannuelsenelec 2015.pdf
7 www.reeep.org/senegal-2014
8 The Emergent Senegal Plan is implemented through the alignment of strategic priorities, sectorial targets, and lines of action, with development projects and programs within a budget framework for the period 2014–2018 (Republic of Senegal, Ministry of Economy, Finance and Planning 2014. Emergent Senegal Plan; Priority Actions Plan 2014–2018. www.finances.gouv.sn/en/Docpdf/PAP_ 2014-2018_of%20PSE.pdf)

References

De Gouvello, C. & Kumar, G. (2007). Output-Based Aid in Senegal – Designing Technology-Neutral Concessions for Rural Electrification. GPOBA, Washington D.C.

ESMAP. (2007). *Maximisation des Retombées de l'Electricité en Zones Rurales, Application au Cas du Sénégal.* Washington, DC: Energy Sector Management Assistance Program, World Bank.

EUEI PDF. (2014). *Mini-grid policy toolkit: Policy and business frameworks for successful mini-grid roll-outs.* Eschborn: European Union Energy Initiative Partnership Dialogue Facility (EUEI PDF).

Fuenfschilling, L. & Truffer, B. (2014). The structuration of socio-technical regimes: Conceptual foundations from institutional theory. *Research Policy,* 43 (4): 772–791.

IEA. (2016). World Energy Outlook: Electricity Access Database, International Energy Agency, Paris. www.worldenergyoutlook.org/resources/energydevelopment/energyaccessdatabase/

IMF. (2012). Regional Economic Outlook. Sub-Saharan Africa. Sustaining Growth amid Global Uncertainty. IMF, Washington D.C. <https://www.imf.org/external/pubs/ft/reo/2012/afr/eng/sreo0412.pdf>

IMF. (2015). *Regional economic outlook Sub-Saharan Africa: Navigating headwinds.* Washington, DC: International Monetary Fund. www.imf.org/external/pubs/ft/reo/2015/afr/eng/pdf/sreo0415.pdf

Lighting Africa. (2012). Policy Report Note: Senegal. www.ecowrex.org/system/files/documents/2012_policy-report-note-senegal_lighting-africa.pdf

Mawhood, R. & Gross, R. (2014). Institutional barriers to a 'perfect' policy: A case study of the Senegalese Rural Electrification Plan. *Energy policy,* 73, 480–490.

Sanoh, A., Parshall, L., Sarr, O. F., Kum, S. & Modi, V. (2012). Local and national electricity planning in Senegal: Scenarios and policies. *Energy for Sustainable Development,* 16: 13–25.

Senelec. (2015). Rapport Annuel 2015. Senelec, Dakhar, Senegal.

Ulsrud, K., Rohracher, H., Winther, T., Muchunku, C. & Palit, D. (2018). Pathways to electricity for all: What makes village-scale solar power successful? *Energy Research and Social Science,* 44: 32–40.

Ulsrud, K., Winther, T., Palit, D. & Rohracher, H. (2015). Village-level solar power in Africa: Accelerating access to electricity services through a socio-technical design in Kenya. *Energy Research & Social Science,* 5: 34–44.

Winther, T. (2008). *The impact of electricity: Development, desires and dilemmas.* Berghahn Books, Oxford, UK.

3 The local context

Socio-technical innovations such as mini-grids and other decentralized infrastructure are not only shaped by situations and processes at the national and international levels, as shown in the previous chapter, but also by the local, socio-cultural context in the places where they are located (Rohracher 2003). Project implementers select locations based on certain criteria, and they design energy models with the local context in mind. Furthermore, the actual functioning of the power provision is also strongly influenced by the encounter between the way the energy system has been designed and the local dynamic situations, including the process of implementing the system in the communities (Ulsrud et al. 2015). This encounter is shaped by the extent to which the project implementers manage to adapt the energy provision to this context, including people's practical needs and economic situations. In turn, the power supply is also likely to influence the socio-cultural context. For these reasons, it is important, before presenting the details of the mini-grid systems themselves, to describe the local contexts within which the case study project implementer aimed to serve the population with electricity from their mini-grids.

Inensus, through their joint venture with Matforce in Senegal, Enersa, implemented their mini-grids in the districts (Senegalese communes) of Méouane and Mérina Dakhar within Thiès. The six sites (with population figures in parenthesis) are in the districts of (1) Méouane, the villages of Sine Moussa Abdou (879), Ndombil (953), Léona and Wakhal Diam (1047) and (2) Mérina Dakhar (695), the villages of Maka Sarr, Sine Lèye Kane, and Mérina Diop. Léona and Wakhal Diam are connected to the same power plant. All the village sites are easily accessible by car from the capital city Dakar and through Thiès city.

Thiès region is located to the east of Dakar region, which is Senegal's smallest and most populated region and includes Senegal's capital city. Thiès region includes the city of Thiès which is the third largest city in Senegal, with about 1.8 million people. Like other regions in Senegal situated away from Dakar, Thiès is prone to inequality, including in public expenditure. For example, approximately a quarter of Senegal's population lives in the Dakar region, yet it absorbs more than half of the country's public resources, including education funds (Kireyev 2013).

The populations of the districts of Méouane and Mérina Dakhar are 35,614 and 32,345 respectively.[1] The two districts are mainly dependent on rain-fed agriculture, with crops that include millet, watermelon, cowpea, cassava, and groundnuts. Agricultural activity in both districts has, however, been hampered by water scarcity, salinization of agricultural land, and natural resource degradation (Lo and Dieng 2015).

Livelihoods, socioeconomic conditions, and political and administrative organization

All the four sites studied are located in villages that are close to each other and have similar socioeconomic and socio-cultural characteristics, including livelihoods, climate, settlement patterns, family organization, and housing.

Settlement patterns are dense, and each settlement covers a small area. Families in the villages are large – several generations live together and polygamy is practiced. From our survey in three of the four village sites, Sine Moussa Abdou, Maka Sarr, and Ndombil, covering about 20% of the households in each village, we found the average number of adults per family to be five women and four men. We did not establish the average number of children because such a question was considered inappropriate. These large extended families are composed of smaller units, consisting of a married man and his wives and children. Boys tend to continue to live at home with their parents after marriage. Each home consists of several rooms (with separate entrances) and often two or more buildings and a fenced compound within which many household chores and activities are done, including cooking and eating. The large family setup in the villages means that one electricity connection could potentially cover a large number of electricity users.

Local livelihoods in the villages are characterized by a high dependence on subsistence agriculture. 57% of households during our quantitative survey responded that subsistence farming was amongst their top two sources of income, as shown in Figure 3.1.

The villages are located in an arid and semi-arid region, receiving limited and unreliable rainfall, typically during three months per year (June/July to August/September). During this short rainy season, the farmers plant drought-resistant crops such as peanuts, millet, and yams, and the harvesting season starts in October. Other sources of income cited in the survey included small businesses such as rope making, tailoring, and grocery

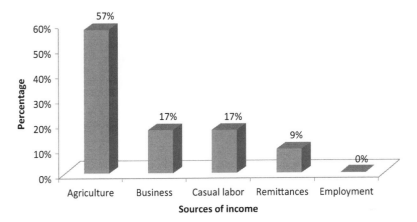

Figure 3.1 Main sources of income as reported by respondents

shops (17%), casual labor (17%), and remittances (9%) received from relatives working outside the villages, mainly Dakar (the capital city) and/or
from the diaspora, notably in France.[2] In addition, many young people, especially men, travel away from the villages in search of employment, especially
during the dry seasons. Livestock and handicrafts are sold at local markets.
The most important market center for all four villages is the market in Ngaye
Mékhé, 15 km away from the village of Sine Moussa Abdou.

During our fieldwork in March 2016, it was observed that women
worked long hours daily on handicrafts such as sewing of large plates
made by straw or doing embroidery, although the payment for this was
very low. The types of economic activities varied slightly between the villages. At the time of the fieldwork, women in Sine Moussa Abdou worked
on embroidery and peeling of peanuts, while in Maka Sarr, rope making
was observed in many compounds (done by the women) and some
peeling of peanuts and rinsing of beans. In Ndombil, several women in
each of the homes we visited were sewing straw plates with intense
effort. Apart from such production in people's homes, a few businesses utilized electricity for income-generating activities, as further explained in
Chapter 6.

A man in Sine Moussa Abdou told us that there is not much work to get
in the village apart from farming. If there could have been someone who
could help us find other jobs, he said, like setting up mills, for example,
that would be very helpful. Another man told us that some special leaves

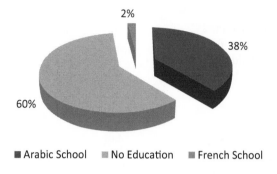

Figure 3.2 Levels of education reported by the respondents (*n* = 55)

grow in this area, which can be used in food, and that they could try to sell them elsewhere.

Large economic differences and wealth disparities between households in the villages were observed during the fieldwork. This was demonstrated through ownership of relatively larger and more expensively furnished houses with assets such as large television sets, furniture, floor carpets, and interior decorations for a few wealthy families, surrounded by more modest houses for the majority of poorer families. Some of the families who lived in these larger houses have relatives in Europe. Our local guide[3] mentioned that these were in countries such as Spain, Italy, and France.

There are two parallel forms of schooling pursued in the villages: the Arabic schooling system that is predominantly religious and the formal curriculum, which is named the French system. Our survey showed that a much higher proportion of respondents (38%) had gone through the Arabic schooling, compared to only 2% who had taken the French curriculum (Figure 3.2). However, the majority of survey respondents reported to have not had any education, reflecting the very high levels of illiteracy in the villages.

There are also marked gender differences in the levels of education amongst adults; men generally have higher levels of education, but predominantly through the Arabic system. In Maka Sarr village, 100% of women respondents reported to have had no education at all, in Ndombil the portion was 67% and in Sine Moussa Abdou it was 75%.

Energy use patterns prior to installation of mini-grids

Common energy sources used before electricity came to these villages through the mini-grids were candles, kerosene lamps, and battery-powered torches for lighting and some solar home systems. A survey carried out before the implementation of Enersa's pilot mini-grid in Sine Moussa Abdou estimated that the average expenditure on energy that could be substituted by electricity from the mini-grid was 12,829.05 XOF (19.5€) per month per household (Peracod and Inensus 2009).[4] We do not know the distribution of expenditure across the population. A common feature of energy expenditure in rural areas in Kenya, as an example, is that the average is above the median figure, i.e., the majority of the households spend below average.

Some villages had a few solar home systems in place before the mini-grids were installed: We observed only one in Sine Moussa Abdou, and in Maka Sarr we saw several old solar home systems when carrying out our interviews and field research.

In Sine Moussa Abdou, a diesel generator had previously been used to run the water pump in the village, while the mini-grid had been used in recent years. According to the men in the village who were responsible for the water pumping, the use of solar power from the mini-grid was less expensive than the use of the diesel generator they used before.

Summary

This chapter has provided a description of the local socio-cultural, material, and spatial context, including livelihoods and economic problems of the four sites where the Enersa mini-grids were implemented. Important characteristics were poverty, social and economic differences between the citizens, short agricultural growing season, long dry season, low literacy among adults, especially women, lack of employment, and struggles to earn money on doing handcraft or small trading with agricultural production.

We acknowledge that the social reality is messy and constantly changing, and that it has many subtle characteristics. We are therefore not claiming to fully understand these communities, far from it. We have, however, gained significant insights crucial to analyze and discuss the socio-technical design of the mini-grid model developed to serve villages bearing the characteristics described above, and how it functioned in practice.

Notes

1 http://senegal.opendataforafrica.org, census data 2013.
2 Remittances do not include money sent by immediate family members, such as a husband or wife, living and working outside their village, although qualitative interviews established that this is common practice, especially with husbands working elsewhere and sending money back to the family.
3 A staff member of the micro-grid project.

4 Based on substitution of candles, kerosene, torch batteries, radio batteries, and expenditure on phone charging.

References

Kireyev, M. A. (2013). *Inclusive Growth and Inequality in Senegal* (No. 13-215). International Monetary Fund.

Lo, H. M. & Dieng, M. (2015). Impact assessment of communicating seasonal climate forecasts in Kaffrine, Diourbel, Louga, Thies and Fatick (Niakhar) regions in Senegal: Final Report for CCAFS West Africa Regional Program.

Peracod & Inensus. (2009). Etude sur le potentiel productif des villages de Sine Moussa Abdou et Sakhor. Rapport provisoire. SEMIS, Dakar, Senegal.

Rohracher, H. (2003). The role of users in the social shaping of environmental technologies. *Innovation: The European Journal of Social Science Research*, 16 (2): 177–192.

Ulsrud, K., Winther, T., Palit, D. & Rohracher, H. (2015). Village-level solar power in Africa: Accelerating access to electricity services through a socio-technical design in Kenya. *Energy Research & Social Science*, 5: 34–44.

4 The socio-technical design

This chapter shows the way that Inensus designed the Enersa mini-grid model and their considerations behind its socio-technical design. The features of the implementation process are also described. This corresponds with the third dimension of our analytical framework (Socio-technical Design). Inensus first came to Senegal in 2007 invited by GIZ. Since wind energy was the initial interest of Inensus, as explained in Chapter 1, they installed five measurement devices for wind speed and conducted surveys in villages. They found that Sine Moussa Abdou would be the best village for a pilot mini-grid project, because of both good wind conditions and potential demand.

After founding Enersa in Senegal with the large Senegalese company Matforce, Inensus could do their pilot, and they had strong expectations to create a profitable activity. They had mobilized some money on their own and they had received contributions from GIZ, who financed parts of the project assets. This combination of own financing and support from GIZ was described by Inensus as a kind of public-private partnership, where GIZ took a role that governments could potentially take. Matforce was a strong technical partner that sold diesel generators, cars, and air conditioning systems and also worked on energy projects.

After implementing the pilot project in the village Sine Moussa Abdou in March 2010, the project implementers held a big inauguration, and many observers were impressed by this innovative, private-sector-led mini-grid project. Enersa now started to look for funding for more villages including co-funding from their side, and got grants and a large loan (1.6 million euros). A "due diligence" study was carried out by a company hired by the investors, and they said, "yes, these guys can do it." Inensus, Matforce, and Enersa managed to raise a combination of debt, equity, and public funds (donor funds) for the implementation of 30 mini-grids, 4 million euros in total. For the young entrepreneurs, all of this was something big, and they were determined to make it a success.

After finally getting the ERIL concession and license contract with MEDER for the pilot project in 2014, after 4 years of work as explained in Chapter 2, Enersa implemented five more projects between July 2014 and March 2015, expecting that the licenses would be given very soon. At the end of 2015, however, the investors pulled out because it was then clear that the regulator would not give further licenses. Legal operation was of course a key requirement for the investors' participation as well as for debt financing.

As expressed with disappointment in 2016 by one of the Inensus founders: "Many people have looked at us and hoped that we could make it work in Senegal. But unfortunately it didn't work, because the regulatory framework was never finalized." This shows the risk of being pioneers who try to break new ground. There was a large contrast between the promising and celebrated pilot and the following regulatory nightmare for the project implementers. It was most likely a difficult and demoralizing experience to handle. However, they went ahead in other places. "This is over now, we have stopped this activity," one of them told us during our joint visit in Senegal. He informed us about their new activities in Tanzania, Nigeria, Sierra Leone, and Myanmar and how they drew on the lessons they had learned through their work in Senegal. Some of this information about Inensus' newest initiatives is included in Chapter 7.

Despite the fact that Inensus' work through Enersa in Senegal got blocked, for reasons that have been thoroughly explained in Chapter 2, the following analysis of the socio-technical design, how it worked in practice and why, as well as the resulting access for the village citizens is still a very relevant source of knowledge for energy start-ups, practitioners, policymakers, and researchers on factors that play a role for sustainability and upscaling of small-scale mini-grids.

The case also provides insights into some of the struggles that take place within niches in the field of rural electrification. The niches consist of spaces that to some extent are protected from competition from more conventional solutions, like the centralized electricity regime based on extension of the main grid. However, there are still plenty of challenges to solve in order to build alternative systems.

The design and implementation process of the mini-grid model

Compared with previous mini-grid models, Inensus' version was more advanced (see Bhattacharyya and Palit 2016; Ulsrud et al. 2011, 2018). When Inensus developed their mini-grid business model, they attempted to find a middle ground between a purely private sector model, where the entire system would be financed using equity and debt, and a public sector model, where public funds (government and/or donor funds) would be used to cover all the capital investment costs. Inensus' motivation for developing a model that included both the private and public sector was based on the following observations:

- The difficulty in successfully developing and implementing purely commercial mini-grid models especially within an unsuitable policy and regulatory framework (where rural electrification is the remit of the government)
- The difficulty of setting affordable end user tariffs in a situation where the entire system is financed using equity and debt
- The difficulty of sustaining mini-grids where the capital costs have been completely subsidized. Systems where generation and distribution have been subsidized are not dynamic and struggle to adapt to increasing demand, because future subsidies are likely to be required to invest in an increase of the generation and distribution capacity, even if the revenue covers day-to-day operation costs and battery replacement.
- The lower chances of success when no private investment is involved, since it reduces the incentives for private companies to ensure long-term viability when operating mini-grids

Inensus' ideas about a kind of private-public financing of their business model were inspired by macro-scale power economy mechanisms in Europe, where legal unbundling (splitting up) of distribution and generation of electricity was implemented. Inensus scaled down this thinking to the village scale, and a central idea behind the model was that a mini-grid has two elements: the fixed assets (the distribution network and power house) and the movable assets (the generation equipment). Private investment is put into the movable assets and public funds are used for the fixed assets. This idea was based on the premise that generation assets like solar PV panels, batteries, and inverters could be easily extended with additional private funds when demand increases, and that these could be more easily relocated to a different site and reused, while this would be difficult with distribution infrastructure. These considerations led to a private sector business model that was based on a private-public partnership financing structure with 60% private financing.

Implementation strategy and electricity services provided

When they selected villages to serve, Enersa consulted with ASER to iden-tify suitable mini-grid sites, following the requirements of the government.

The main factors considered were distance from the grid and electrification plans. Typically, villages more than 5–10 km from the grid were targeted by off-grid solutions. Senegal's concessional framework, that allocates regions to private sector to implement electrification activities as explained in Chapter 2, also had to be taken into account. Either sites outside these concessions would be considered or an agreement reached with a concessionaire and ASER to work in a given concession area. The villages considered by Enersa were outside the already allocated concession areas. Once a village had been selected, a cooperation partnership with ASER was signed. Enersa thereafter applied for the required licenses from MEDER, which led to the complicated process explained in Chapter 2.

Before making the final decisions about where to locate their mini-grids, Enersa considered the suitability of an identified site by undertaking what they called a macro and micro survey, after an introductory meeting with the local leader/government representative in the village.

The macro survey was to:

- Collect socioeconomic information, e.g., population statistics, economic activity (income sources and patterns), potential for economic growth, etc.
- Determine the number and type of different potential customers, e.g., households, businesses, institutions, social services (e.g., water supply, health facilities, educational institutions, churches/mosques)
- Determine accessibility, e.g., to neighboring un-electrified villages (that can form a cluster or be interconnected), fuel stations (supply of diesel for the backup generator), and telecommunication services, which might be suitable for connection as anchor customers[1]
- Identify potential supporting partners, e.g., NGOs and micro finance institutions, who could help provide consumer finance for purchase of electrical appliances for productive use/income generation

The micro survey was to determine the potential electricity demand in the village. This information was used to design and size the generation and distribution system by use of Homer software for both initial demand and anticipated future demand. Such assessment of the demand is crucial according to our Inensus informant. During the micro survey, Enersa staff was accompanied by a person from the village.

Inensus had started with technical expertise, but found that they had to develop their skills in several directions, such as business management and how to work with communities. When working on the Enersa mini-grids, they were committed to prevent conflicts in the villages about the electricity system, and they told us that small mistakes can make big problems. They would always work with the village leaders, since the decision-making structures in the villages were not transparent. They also emphasized the importance of working with the village leaders as equal partners.

This mini-grid model sold electricity through units called Block units. Based on the tariff model approved by CRSE for the pilot project (see Chapter 2), each Block unit was equivalent to 1.4 kWh of electricity per week with the power that could be drawn at any given instant capped at 50 W. The available energy could be used for up to 1 week, after which the Block would expire (even if all the available energy hadn't been used). Customers could subscribe to using as many Blocks per week as they wanted, up to a power cap of 2.3 kW. In Chapter 6, we provide details on appliances possible within a given number of Blocks. In addition to the Block energy, people could purchase "extra energy" in the form of electricity units (kWh) without limitation of power. This way, businesses could have the possibility to use higher power loads, and users would have an opportunity to bridge times when they have exhausted the Block energy before the end of the week or cover short peaks and additional energy demands. The "extra energy" wouldn't expire, since it didn't have a time limit. Such flexible energy was sold at higher unit prices. In addition, there was also a business tariff, based on actual consumption, not on Blocks. This tariff was much lower, about 1/3 of the household tariff.

Based on their projected electricity demand and ability to pay, interested customers subscribed to using a certain number of Blocks per week. To guide them through this process, they received information about how many Blocks they would need in order to power different types of appliances. The project implementer also informed about the conditions of connecting as well as regulations for usage of electricity. The information activities included visits in the households and public meetings. Once they subscribed to using a certain number of Blocks per week, customers signed an expression of interest for electricity and then paid for internal wiring and connection.

Customers were required to make an annual commitment to purchase the subscribed number of Blocks every week. Although not strictly enforced in practice, the project implementers found this to be necessary because they had to make an upfront investment in electricity generation and distribution capacity to provide for the electricity needs of the interested customer. If the customer would not buy their subscribed portion of generated electricity, then the investment for this connection would go to waste unless another customer took it up.

The technical system design and the investment costs

The typical load planned for in each of the Enersa mini-grids, as established through the micro survey, was 5 kW peak load and 20 kWh of daily energy for subscribed customers and an additional 0.5 kW peak load and 2 kWh of daily energy for street lighting. A 10% contingency was built in, bringing the total design load to 6.05 kW and daily design energy demand 24.2 kWh/day. There were small differences in the population size between the selected villages. In March 2016, the four grids covered by this study served the following number of connections: Sine Moussa Abdou 76, Maka Sarr 61, Léona and Wakhal Diam 94, and Ndombil 65.

Generation was mainly solar PV (10 kWp) with battery storage (2,220 Ah/48V, approximately 70 kWh of energy storage capacity[2]) and a backup generator (17–33 kVA). An additional 5 kW wind system was a considered option where there are good wind speeds.[3] The battery inverters were either 10 kW – single phase or 30 kW – three phase (10 kW per phase). The generator was configured to start automatically when the battery state-of-charge would get to 50%.

An AC bus system configuration was used as shown in Figure 6.1, i.e., the interface between the various components was AC; the PV array output was converted to AC with an inverter (or a number of string inverters) and the battery was connected to the other components via an inverter charger. The wind system could also be connected to the AC bus, but in the first mini-grid (Sine Moussa Abdou), it was connected to the battery via a voltage controller/battery charger.

A power station to house the batteries, inverters, balance of system components, and backup generator was also incorporated. The power station was typically designed to maximize cross ventilation and keep insects out.

In some of the mini-grids, there was overhead electricity distribution (aerial bundled cable) and in others the cables were underground. The communities provided land and wayleaves required for the power house and the distribution network in kind. Each of the distribution networks was about 1.5 km long, with a line voltage of 400 V. Inensus also considered what would happen in case the main grid would arrive later. They made the technical setup of the mini-grids "grid-compliant" by using grid-tie inverters, and in such a case the power station could be used to feed electricity into the grid. This could then help to stabilize the voltage in the main grid.

A key technical innovation in Inensus' business model was the prepaid metering system they had designed. It had the following features:

• The smart meters mentioned above (one for every three separately metered connections) would automatically disconnect customers when they exceeded their power or reached their energy limit (as defined by their subscribed electricity Block). The digital display of these meters, called "micro-power meters," showed users how much electricity they had left. Inensus chose to combine three connections to one meter and to mount them in public because this would make control easier and discourage tampering with the meter. Neighbors would react if somebody tried to tamper with it. According to Inensus, three customers per meter was suitable, because a higher number would result in too many and too long cables.

• A console for each mini-grid and one chip card for each customer. When a customer paid, the console was used to transfer the payment information to the chip card. The card was inserted thereafter into the smart meter so that the information could be transferred to the meter. The customer could then access the Block of electricity paid for.

Streetlights had been installed in all the villages, connected to a single meter. All customers were supposed to contribute to paying for the electricity for these lights.

Operational design

In addition to procuring equipment, installing, and commissioning the mini-grids, Enersa were also involved in undertaking internal wiring for electricity customers, though a household was free to get the wiring done by other electricians. Internal wiring for one electricity Block typically provided for three power outlets, five indoor lights (8 W), and one outdoor light (11 W), with Enersa paying for the costs upfront, 50€ per block (50 W).

The electricity tariff that consumers would pay for electricity from the Enersa mini-grids comprised:

• Electricity Block payment – 1.4€ per week, i.e., 5.7€ per Block per month (equivalent to 1€/kWh)
• An upfront 50% payment for the investment in their indoor electrical installations. This is about three monthly payments of the respective amount the customer has signed up for.
• Installment payments for the rest of the cost of internal wiring which is paid over 10 years (at 15% interest rate) – typically an additional 8–14% of the monthly electricity payment.
• For extra energy, the price was 1.5€ per kWh.

With time, the work on internal wiring in the customers' buildings was taken up by village electricians, especially for new customers on operational micro-grids. These electricians had been trained by Enersa. Under this arrangement, customers paid Enersa for the upfront contribution and then made their own arrangements with the village electrician for internal wiring.

Operational roles were divided as follows:

a   Enersa – the mini-grid owner and operator

- Set the tariff and undertake monthly visits (from their office in Thiès town) to collect electricity payments from the local power plant operator
- Advanced system maintenance (of batteries and generators) and replacement of system components
- Connecting new customers
- Monitoring system performance directly and through the local operator
- Engaging with the customers via the village electricity committee to address customer and operator issues related to the electricity system

b   Local power plant staff (or seller, as explained above, a person who is a resident of the village)

- Basic operation and maintenance (i.e., cleaning modules, filling the diesel tank, turning on the backup generator, monitoring system performance indicators (e.g., battery state of charge), and reporting problems to Enersa).
- Collecting electricity payments and loading electricity units onto the chip card
- Collecting applications for new connections and sharing this with Enersa
- Receiving an 8% commission on electricity payments received and a small additional payment for provision of maintenance services such as cleaning the solar panels

c   Village electrician

- Installing internal wiring for customers (mainly new/additional customers or existing customers on operational grids)
- Troubleshooting and maintenance of internal wiring

d   Village electricity committee

- Group of village representatives, not paid for their work, responsible for collecting customer feedback and/or issues facing electricity customers to be raised and resolved with Enersa

• Owning the fixed assets (the power house and the grid) which cannot be taken away, since this has been paid for through grants and is seen as public goods

This collaboration with the village electricity committee was crucial and, in each village, a contract was signed between Enersa and the village, through the leader of the village. An initial idea for the collaboration with the villages was that the village electricity committee would purchase the electricity in bulk and sell to people in the village. Another idea was to have an agent or entrepreneur in such a role. This was not implemented because the project implementers found that in the villages targeted, the communities did not possess the skill sets necessary to take over responsibilities of ownership and management of the distribution grid.

At the time of our fieldwork, the company Enersa had five staff members, all of them from Senegal. The traveling distance from the office to the villages was one to two hours by car. The company had an office in the building of the Inensus partner for this joint venture company, Matforce, in the city Thies.

Summary

The three young graduates who started Inensus developed a sophisticated and innovative "socio-technical design" for a private-sector-led, public-private-partnership type of mini-grid model and implemented it through the joint venture they started with Matforce in Senegal, Enersa. This chapter has presented its details as well as several of the underlying considerations for the project implementers' choices of solutions. Among the central visions that guided the project was that private companies could play an important role in providing electricity access where the public sector is not able to provide such services, at the same time as acknowledging that this would be difficult for a purely commercial model. Thus, a public or donor contribution for the investment costs (40%, for the fixed assets) was included. Inensus also emphasized good cooperation with the communities and adaptation of the model to the local conditions and needs. They tried to balance this with the necessity of economic profitability, which they saw as inevitable. Thus, their tariff per kWh was higher than the national utility Senelec's. The following chapter looks at how this thoroughly developed model functioned in practice and what was learned.

Notes

1 An anchor customer is a significantly larger customer than the majority of customers, having a constant demand for electricity and thereby paying a much

higher amount per month for electricity than normal customers. This helps in securing sufficient revenue for the mini-grid operator, but also creates vulnerability in case the anchor customer disappears.
2　Assuming 75% depth of discharge and 90% battery efficiency.
3　Inensus developed a wind resource assessment/measurement system suited for small wind projects.

References

Bhattacharyya, S. C. & Palit, D. (2016). Mini-grid based off-grid electrification to enhance electricity access in developing countries: What policies may be required? *Energy Policy*, 94: 166–178.

Ulsrud, K., Rohracher, H., Winther, T., Muchunku, C. & Palit, D. (2018). Pathways to electricity for all: What makes village-scale solar power successful? *Energy Research and Social Science*, 44: 32–40.

Ulsrud, K., Winther, T., Palit, D., Rohracher, H. & Sandgren, J. (2011). The Solar Transitions research on solar mini-grids in India: Learning from local cases of innovative socio-technical systems. *Energy for Sustainable Development*, 15 (3): 293–303.

5 Findings on how the model functioned in practice and why

This chapter and the next show what happened when the technical innovations and smart business model met the social reality. As explained in Chapter 1, we aim at in-depth *understanding* of what happened and why, as seen from the perspective of different people involved. In addition, we suggested a small element of evaluation, or rather assessment, of the functioning of the mini-grid systems, because this helps in understanding how the operations worked and why. Some desirable qualities of an electricity system are the following, as explained in Chapter 1: The system should provide access to electricity for most of the population or ideally all, and the system should be possible to sustain economically and operationally, including a sustainable model for ownership and management. We also suggested that an electricity system should be sensitive to gender and to the societal context, and that it should be possible to replicate/scale up the electricity provision. Finally, we mentioned the importance of analyzing the qualities of the electricity access. All these aspects are analyzed in this and the following two chapters.

Unexpected and unintended outcomes are inevitable

According to studies of social and technological change, unexpected and unintended outcomes of innovative strategies for change are inevitable, and important learning processes are created when trying out new solutions in practice (Williams and Sørensen 2002). It is therefore interesting to look at such outcomes in the Enersa case. The actual functioning of the Enersa power supply was near to the initial expectations of the implementers in several ways, while there were also differences.

Operational sustainability – and factors influencing it

Operational sustainability is a necessary quality of all energy projects, and according to previous research, and expectedly, this is an easier achieved goal than economic sustainability, affordability, and other qualities of an energy system. Operational sustainability can be defined as practical, well-functioning operation and maintenance going over a long time (many years – until the system is replaced by something else). This includes good and manageable daily routines for power plant operators and other people involved and relevant skills among these. It also includes good systems for repair, access to spare parts and expertise when needed, well-functioning structures for ownership and decision-making, well-functioning technical equipment, and good systems for informing, serving, and managing different types of customers so that they can optimize their benefits from the electricity services provided (Bhattacharyya and Palit 2014; Ahlborg and Sjöstedt 2015). With good advice, the customers can get more out of their tariff payments in a system like Enersa's, where customers can choose the types of subscriptions that suit them best. However, operational sustainability has different degrees – here we have described an optimal situation, while operations might keep going without achieving all these qualities.

In Chapter 4, we explained how the operation of the Enersa mini-grids was planned to work, including a range of roles intended to be filled by several different actors. We found that these operational roles functioned in similar ways to how they had been planned. There were several reasons for this, including willingness to make small adjustments and improvements after implementation. One important factor for operational sustainability was that the operation of the power plants was mostly taken care of by Enersa's own staff, ensuring regular maintenance like cleaning of PV panels, replacing batteries, and maintaining generators. The few tasks of the local sellers/operators, therefore, seemed unproblematic for them to do. An important factor for the performance was also technical equipment of good quality with advanced functions, which mostly worked well. The role of the sellers seemed to be much more important than the role of the village electricity committees, and we got the impression that people went to the seller if they needed information or had problems. In Ndombil village, we were told that in case there was a big problem for the whole village, the community members would come together to discuss, but that people mainly would go to the seller individually. Enersa visited the villages once per month, and this was important in order to support the local sellers and follow up on the problems experienced locally.

Political factors at the local level have played a role in previous cases of village-scale electricity systems (Ahlborg 2017; Ulsrud 2015). We found a few small signs of such factors in the Enersa case. Some men said that the people living on the same side of the village as the seller (in Sine Moussa Abdou) received more information than did people on the other side. One informant in Ndombil said that the grid was recently expanded more on the side where the village leader lived. However, according to our data, such political factors at the local level were not among the main factors that influenced the functioning and access in this case.

The prepayment of electricity tariffs and automatic control of the consumption was also important for the mini-grid performance. These solutions solved several typical problems in the customer management of previous mini-grids, such as customers failing to pay, unmetered supply, overload of the battery banks, and deep discharging of the batteries, which occurred in the pioneering solar mini-grids of West Bengal Renewable Energy Development Agency (WBREDA), starting in 1996 (see Ulsrud et al. 2011). In previous cases, problems in this field often reduced the ability to do major maintenance, like battery replacement, or to invest in expansion of the power generation, and the users ended up losing their electricity access or having poor-quality access. There were operational problems related to the smart cards to be used by the customers, but this is further explained in the next chapter, where we focus on the perspective of the people in the villages.

One of the challenges with the weekly Block system identified during our fieldwork was that it did not currently account for power outages. In one of the mini-grids, because of the failure of one of the solar PV inverters, the system was operating at half its capacity and therefore the backup generator had to be run frequently. As a result, delays in purchasing fuel sometimes resulted in up to day-long outages. When this happened, customers lost out as the system did not enable them to recoup hours or days lost as a result of power outages. The customers' perspective on the Block system is further elaborated in Chapter 6.

Many lessons had been learned over the years through the practical experience from the pilot project in Sine Moussa Abdou. "We have learned a lot from this village, so we do not have the same problems in the other villages," an Enersa staff member told us. He further explained that one of the mistakes they made in the beginning was that they let people have free power at first so that they could get experience with it. This made them less willing to pay later and less satisfied.

Some lessons were also learned from one power plant to the next on technical solutions such as the size and design of the power plant building (the first one was larger than necessary and some of the following were too small), including ventilation (natural ventilation through a good design of the power plant house prevented overheating, but bats and insects should be kept out), the features of the electricity grids (there were pros and cons of both underground and overhead grid lines), the designs and materials for the solar module mounts (locally produced versus imported), the sizing of the power plants, and projection of demand (smaller than indicated by the surveys).

There was no streetlight to be seen in the evenings. Enersa did not switch them on because they did not receive any payment for this. Payment for this depended on self-organization and joint decisions on payment from the community members, and this seemed to be difficult to solve. The main problem was people's different willingness and ability to pay. In Ndombil, for example, one of the key informants told us that the women were very interested in turning on the streetlights. One of the women we interviewed told us that she really wanted to have it on. She found the village to be too dark, and she was afraid of going out in the dark. She also said that "when the street lights are on the village is really beautiful." Another woman said that it is important during the rainy season because there is grass and weeds in the streets, so when children walk there in the evening there might be snakes that they cannot see. The chief told us that people had suggested that an amount be included in each power bill for streetlights, but that the company said no. The company later told us that such an arrangement is not permitted, as the tariff model calculated by CRSE does not include the cost for public lighting, and they were not allowed to increase the tariff. The people tried to create a system for contributions from each household, but not everyone was willing or able to contribute. In areas with electricity from the main grid, the communes (districts in Senegal) normally pay for streetlight, but in the mini-grid villages it had not yet succeeded to get such an arrangement.

Economic sustainability – the economic performance and its reasons

Creating an electricity system that can sustain itself economically and even generate a profit is the target that seems most difficult to reach in previous mini-grid projects and other village-level electricity supply. This target is not easily achievable for any kind of rural electrification, not even the sale of or provision of services from solar products to those people who have a relatively high ability to pay for this (Alzola et al. 2009; Camblong et al. 2009; Ahlborg and Sjöstedt 2015; Palit and Chaurey 2011; Muchunku et al. 2018; Ulsrud et al. 2018).

An electricity system is able to sustain itself economically if it generates enough revenue to cover all necessary costs of operating and maintaining the local power plant in the long-term perspective. A potential surplus could be saved for expansion so that it reaches out to larger areas and more people. In addition to economic sustainability of the operation and maintenance of the mini-grid, cost recovery is another ambition, and for the Enersa model, partial cost recovery was the aim, while grants covered 40% of the investment costs. Enersa and its owners, Inensus and Matforce, were aiming at economic sustainability and profitability. They aimed at recovering their investment costs, covering the costs of operation, maintenance, and expansion for the future, as well as covering the costs of debt financing and allowing for generation of profits. Their concept was to do this based on purely commercial principles, after the investment in the distribution grid and power house had been covered by donors or potentially the public sector.

The budget for cost recovery and expenses for operation and maintenance had been calculated based on the estimated revenue from 30 power plants (the planned target). This would also cover overhead costs like offices, equipment, and Enersa staff. The current staff (five persons) had been hired based on the planned upscaling to 30 villages, and the costs were therefore higher for the time being than what six power plants could support. Enersa was fully aware of this and was dealing with the situation in different ways. They did consultancy on project work for others, for instance, based on their areas of expertise and experience on mini-grids.

Battery replacement had not yet been necessary when we did our fieldwork. Most of the batteries were new, due to the short time since the power plant installation, but even the batteries installed 7 years ago in Sine Moussa Abdou were still working without problems. The long battery life in Sine Moussa Abdou most likely resulted from a sophisticated technical solution for battery management and frequent maintenance.

When it comes to the details of Enersa's economic performance, information was not shared with us, and this would not be natural for a private sector company to do. However, our informant from Inensus told us about the economic challenges of operating small electricity systems. One challenge is that the smaller the electricity systems are, the more difficult it is to achieve scale benefits (economics of scale). Moreover, people's electricity use in the households gives a peak in the evening, and the system has to be dimensioned for this. During the day, therefore, the power is not fully utilized. In larger communities, there is more use of electricity during the day because of more businesses and public customers. In other words, there is a more balanced load profile, and this gives more profitable operation. Enersa hoped that new business activities would be created in the villages based on electricity, and we discuss the challenges of starting such new income-generating activities in Chapter 6.

After implementation, the demand for electricity developed differently in villages that seemed to be similar in size and economic activity. In a village we did not study, Sine Lèye Kane, the demand had grown much slower than expected, while in Maka Sarr and Ndombil it had grown faster than expected. Due to the fast-growing demand in Ndombil, where the current solar PV capacity was 5 kW, the diesel generator was running much more than in the other villages. Enersa planned to take 5 kW of solar modules from Sine Lèye Kane (and thereby reduce that power plant to 5 kW) and install them in Ndombil. In addition, 5 kW of new solar modules would be installed, to get 15 kW in total.

Operation and transaction costs are much higher per kWh for small utilities like this than for larger utilities. Enersa worked on the reduction of such costs (the costs of using cars, offices, work hours, etc.). At the same time, they also felt it is important to create understanding for the costs of operating mini-grids among financers.

Summary

Inensus' socio-technical design led to mini-grids that mostly functioned well in practice – they were operationally sustainable. This was mainly due to good-quality equipment, systematic routines for maintenance by company staff, and suitable tasks and training for local operators and committees. The regular follow-up visits (monthly) to each of the villages were also an important part of the routines. On the economic performance, one of the lessons learned by Enersa and Inensus was that there would be better chances to make business in larger villages where there was more economic activity from the outset of electrification. The challenge of economic sustainability is further discussed in the following chapters, in relation both to the users' and potential users' perspectives and situation and to challenges of upscaling/replication.

References

Ahlborg, H. (2017). Towards a conceptualization of power in energy transitions. *Environmental Innovation and Societal Transitions.* http://dx.doi.org/10.1016. 2017.01.004

Ahlborg, H. & Sjöstedt, M. (2015). Small-scale hydropower in Africa: Socio-technical designs for renewable energy in Tanzanian villages. *Energy Research and Social Science*, 5.

Alzola, J. A., Vechiu, I., Camblong, H., Santos, M., Sall, M. & Sow, G. (2009). Microgrids project, Part 2: Design of an electrification kit with high content of renewable energy sources in Senegal. *Renewable Energy*, 34 (10): 2151–2159.

Bhattacharyya, S. C. & Palit, D. (2014). *Mini-grids for rural electrification in developing countries: Analysis and case studies from South Asia.* Switzerland: Springer.

Camblong, H., Sarr, J., Niang, A. T., Curea, O., Alzola, J. A., Sylla, E. H. & Santos, M. (2009). Micro-grids project, Part 1: Analysis of rural electrification with high content of renewable energy sources in Senegal. *Renewable Energy*, 34 (10): 2141–2150.

Muchunku, C., Ulsrud, K., Palit, D. & Jonker-Klunne, W. (2018). Diffusion of solar in East Africa: What can be learned from private sector delivery models? *WIREs Energy Environ.* https://doi.org/10.1002/wene.282

Palit, D. & Chaurey, A. (2011). Off-grid rural electrification experiences from South Asia: Status and best practices. *Energy for Sustainable Development*, 15 (3): 266–276.

Ulsrud, K. (2015). Village-level solar power in practice: Transfer of socio-technical innovations between India and Kenya. *PhD dissertation*. Department of Sociology and Human Geography, Faculty of Social Sciences, University of Oslo.

Ulsrud, K., Rohracher, H., Winther, T., Muchunku, C. & Palit, D. (2018). Pathways to electricity for all: What makes village-scale solar power successful? *Energy Research and Social Science*, 44: 32–40.

Ulsrud, K., Winther, T., Palit, D., Rohracher, H. & Sandgren, J. (2011). The Solar Transitions research on solar mini-grids in India: Learning from local cases of innovative socio-technical systems. *Energy for Sustainable Development*, 15 (3): 293–303.

Williams, R. & Sørensen, K. H. (eds.). (2002). *Shaping technology, guiding policy: Concepts, spaces and tools*. Cheltenham: Edward Elgar. 404 pp.

6 Resulting access to electricity and the perspectives and experiences of the people in the villages

In this chapter, we focus on the users' perspectives, as well as the perspectives of non-users of the electricity from the Enersa mini-grids. Following the analytical framework presented in Chapter 1, the dimension analyzed in this chapter of our book is the question of how the electricity system studied in Senegal worked for the people who needed it, including how accessible, affordable, and reliable it was, as well as the reasons for this situation, which may change over time. We investigate who was able to connect to the electricity supply, what people were able to use the electricity for, at what times, and why this was so. Such analysis is important in order to understand the factors that influenced people's ability to gain access to electricity services and stay connected, as well as the extent to which the electricity system (the way it had been designed and the way it worked in practice) met people's needs for electricity services over time.

This chapter has six sections. The first section provides a description of the ways in which electricity was being used in the four study villages by various types of consumers. The following five sections discuss various factors that influenced people's opportunities to gain access to and use the services in ways that suited their needs, economic situations, everyday lives, and practices. As part of this analysis, we discuss how the electricity system suited the local context in which it had been implemented.

The actual use of electricity services

One challenge involved in electrification is that even if electricity is made available in a place, many people may still not be able to benefit. In many cases, significant portions of the population may still not obtain access (Palit and Chaurey 2011; Ulsrud 2015). However, in the four villages studied here, it was clear that broad access to basic electricity services had been achieved, even in villages in which less than 1 year had passed since the start-up of the power plant. In our survey, which covered about

20% of the households[1] in three of the villages (in Maka Sarr, Ndombil, and Sine Moussa Abdou), 92.5% of the respondents had connections, even though we actively tried to find respondents without connections. This connection rate is high compared with typical electrification rates (the portion of people connected to the electricity supply in each village) in areas covered by the national grid in Sub-Saharan Africa. The connection rate in the studied villages was also high compared with what is typical for solar home systems and solar lanterns in terms of the portion of households having such systems in a given location, as found, for instance, in a survey carried out in Kitui in Eastern Kenya in 2015, as well as in Homabay in Western Kenya (Winther et al. 2018), which has one of the highest densities of such stand-alone systems.

Electricity use in households

The households in the four villages mainly used electricity for lighting (in the evening, at night, and in the early morning), phone charging, radio, fans, and TVs, according to our qualitative interviews and quantitative survey. Computers were also mentioned. One customer, for instance, told us that "we used to buy candles for the children to study. Now, we have light, TV, and radio." Her family used light for about 5 hours at night and a small TV for 4–5 hours per day, but she felt they needed more electricity to watch TV in the afternoon rather than only in the evening.

The quantitative survey provided an overview of the portions of power plant customers who used electricity for various purposes, as shown in Figure 6.1. Outdoor lighting was used by 90% of customers, and indoor lighting was used by 76% of the customers included in the survey. Eighty-seven percent charged mobile phones, 59% used electricity for radio, 41% used it for TV, and 20% used it for fans. Freezers and fridges were each used by 9% of customers.

People showed a strong interest in using electricity because they found it to be useful in their lives. During the qualitative interviews, as well as when we were thanking people for answering the survey questionnaire, a significant portion of the respondents gave positive statements about the benefits of the electricity provided. This often came up of people's own initiative. Other times, it was a response to our question, "what is good about electricity?" The answers to this question included "power is easier to use than candles and kerosene," "we used to charge phones in another village before," and "we are really happy about the power because now, we can watch TV and see what happens in the world." One further comment was "whenever the cell phone is off, we can charge it." Due to the use of smartphones, on which people watched videos and used Facebook and

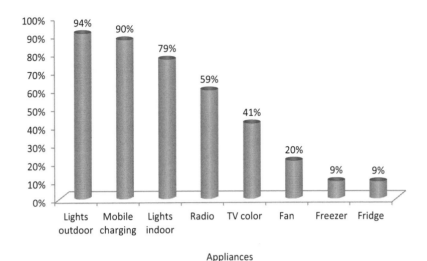

Figure 6.1 End user applications among power plant customer respondents

WhatsApp, it was especially important for people to have easy access to charging. Another woman told us that TV, lights, the radio, and phones were important to her and that "if one is without electricity for one day, one becomes blind." When we asked her whether TV or light was most important to her, she said that it was the light "because if you have only TV, you cannot see when you work." The light in her home was usually on from 6:30 or 7:00 p.m. until midnight. When we asked a group of women about how TV affects their lives, they said that they learn about breaking news and watch movies and religious broadcasting, as well as that their children receive education in this way. One sick man told us how important it is for him to use a fan because the weather is sometimes very hot. "Then, I switch on the fan, and I feel cooler. This is really good for me," he said.

Based on this interest in using electricity for various purposes, people aspired to use more than what was financially possible for them. In addition, among those who did not have a functioning TV at home, it was common to watch TV in other people's homes. People who did not have a connection charged their phones at neighbors' houses. Because some people used their phones as torches, they also obtained some light in this way. Moreover, in Sine Moussa Abdou, the joint water supply for the village was operated using electricity from the mini-grid. A diesel generator was used previously, and this was more expensive. The electric water pump connected to the solar power plant was switched on for 7 hours during the day to fill up the elevated water tank. The water was then sufficient to supply the households for 2–3 days.

The use of electricity among the customers varied according to the type of connection, or Block, people had subscribed to. As mentioned in Chapter 4, Block 1 allowed for five compact fluorescent lights (CFL) of 8–10 W each, radio and phone charging, and two sockets, which amount to a maximum capacity of 50 W. This could be used for up to 4 hours per day on average, which was the same for all Blocks. For Block 2, the appliances allowed were seven lamps, mobile charging, and a small TV, as well as three sockets. Block 3 allowed for 12 lamps, phone charging, a larger TV, a fan, and four sockets, with up to 150 W switched on at once. Finally, Block 4 could power 14 lamps, mobile charging, a larger TV, two fans, and five sockets – up to a 200 W capacity. In addition, people could purchase "extra power" if they wanted to add larger loads or extend the time electricity was available.

Table 6.1 shows the distribution of Block options among the survey respondents. Because the households were permitted to revise their subscriptions once per year by either downgrading or upgrading across the Block options, the table shows changes over time.

Twenty-eight percent of the connected respondents had subscribed to Block 1, 48% had subscribed to Block 2, 20% had subscribed to Block 3, and 4% had subscribed to Block 4. The percentage of unconnected respondents was 7.5%. Thus, almost half of the households included in the survey had chosen Block 2.

The Block framework can be compared with the energy access multi-tiered framework used by the UN Sustainable Energy for All (SE4All) initiative. It has been developed by the World Bank/ESMAP and has now been further developed into a hierarchy of Energy Access Indices according to ESMAP's Conceptualization Report on how energy access can be

Table 6.1 Household energy choices by power Block

Blocks subscribed to				
Block	*Initial subscription*		*Current subscription (March 2016)*	
	Count	*%*	*Count*	*%*
Block 1, 4 h/d, 50 W	10	22%	13	28%
Block 2, 4 h/d, 100 W	21	45.5%	22	48%
Block 3, 4 h/d, 150 W	12	26%	9	20%
Block 4, 4 h/d, 200 W	3	6.5%	2	4%
Total	**46**	**100%**	**46**	**100%**

Table 6.2 Multi-tier matrix for access to household electricity services (ESMAP 2015)

Tier 0	Tier 1	Tier 2	Tier 3	Tier 4	Tier 5
	Task lighting Phone charging	General lighting Television Fan (if needed)	Tier 2 AND any medium-power appliances	Tier 3 AND any high-power appliances	Tier 4 AND any very high-power appliances

defined (ESMAP 2015). Table 6.2 describes the various energy service tiers used by the UN.

Blocks 1, 2, and 3 would enable the Tier 2 level of service (between 200 and 1,000 Wh/day), while Block 4 would provide Tier 3 service (a minimum of 1.0 kWh per day). The energy services used by most households in the three study villages are predominately in Tier 2 and, to a limited extent, Tier 3. In Tier 2, we have both indoor and outdoor lighting, phone charging, fans, and some entertainment in the form of radio and color TV. In this context, only two types of appliances – fridges and freezers – would be classified as medium-power applications (Tier 3).

However, although this might look like a relatively low-level of electricity use, having previously performed research in areas where the majority of the population struggle to be able to have even one electric light in the form of a rented or purchased portable lantern (see Ulsrud et al. 2015), it was striking to see all the lights and other appliances used in the households in these villages, even those within Blocks 1 and 2.

Income generation via electricity use

After analyzing the use of electricity in households, we turn our attention to the use of electricity as a part of business. We are aware that non-commercial uses of electricity can have economic dimensions (Cabraal et al. 1996), but here, we choose to focus on the direct use of electricity in commercial activities. It is important to be aware that businesses and households are often located together, and this was the case in these villages, according to the power company. Sometimes, a home had two connections – one household connection and one business connection – but other times, a home had only a business connection, which also supplied the household. The commercial (or business) tariff was lower than the household tariff, 325 XOF (30€¢)/kWh per week for businesses and 925 (1€¢)/kWh) for households, because the household tariff is very high for large consumers. Enersa was piloting a post-pay metering system for commercial customers.

The same survey respondents described above were asked whether they used electricity to generate income. Three respondents in Maka Sarr used electricity for ice and ice-cream businesses, one used it for light for a business, and one used it for tailoring. One way of using the ice was for chilling drinking water in this very hot climate. In the qualitative interviews, we were told that eight women in total were engaged in freezer businesses in Maka Sarr. In Ndombil, one respondent had an ice and ice-cream business and used light for this business. In Sine Moussa Abdou, none of the respondents used electricity for business purposes.

During our fieldwork, we also observed a milling business in Ndombil. Commonly available milling machines have relatively high power ratings (7.5 kW, 3-phase) and therefore cannot be connected. Enersa had been investigating the possibility of retrofitting a motor soft starter to enable mills to operate on the mini-grid. In addition, on one of the mini-grids, two machines (2.2 kW and 1.5 kW) were used for grinding coarse and fine powder in an enterprise connected with the mini-grid. Both were run for one hour per day during daytime, and when one was running, the other was kept non-operational. The unit paid a fixed tariff per month of 35,000 XOF (53.4€) as of February 2016 (this was the first month of operation). Total earnings in February were approximately 55,000 XOF (83.8€). The purchase cost of the machine was 750,000 XOF (1,143.4€), of which 500,000 XOF (762.2€) was paid by the owner's son, who was working in Spain.

For fridges, four or more Blocks were required. Two customers we met in Maka Sarr who used fridges paid 25–30€ monthly for electricity.

There was some interest in engaging in milling, pumping water for irrigation in order to grow agricultural products outside the single rainy season, refrigeration, and use of freezers. Pumping water for irrigation is difficult because farms are at a distance from the mini-grids. Moreover, people in the various villages explained to us that it was difficult for the people there to invest in new businesses that could benefit from electricity, because the villages had little economic activity other than agriculture and handicrafts.

There were differences between villages, despite their being in a very similar area and within short distances of one another. Maka Sarr and Ndombil were slightly larger villages than Sine Moussa Abdou and had more economic activities both with and without the use of electricity. In Maka Sarr, one seasonal activity that occurred during our fieldwork was rope making, which was done in many compounds via the use of hand-winded machines to twist the ropes. This provided daily earnings of about 1050 XOF (1.6€) per day.

In Sine Moussa Abdou, people were interested in starting businesses with electricity, but they saw it as difficult. After sitting down with a group of women who had gathered in their meeting place in Sine Moussa Abdou, as well as a few passing men, we asked them whether electricity allowed any income-generating activity. One woman answered as follows:

> Power is good, but I want the guys to reduce the price so that I can buy more power to get income. If there is no harvest, and there are harvests only some times in the year, there is no income. I want lower prices so that I can buy a freezer, and I want to get more water to grow potatoes and vegetables in the fields by using electricity to pump the water.

We later learned that a lower tariff was offered for freezers than for the Blocks, but this woman probably still found it to be too high. She added that "now, everybody is peeling peanuts, and nobody buys." A woman interviewed elsewhere in the village expressed that "there is lack of jobs. That's why we don't have money." The same kind of message was delivered by many of the informants. Some people in Sine Moussa Abdou and Maka Sarr asked our research team whether we could create a project that would generate jobs. Some said, "we are ready to work in any project because we are committed." Others were more direct and said, "Those who made efforts to bring electricity should find solutions, bring refrigerators and freezers, and help with operations."

One tailor in Sine Moussa Abdou had previously used an electric sewing machine and thereby increased his production and income. However, the demand for clothes in the area varied with celebrations and festivals, and it was difficult to pay the electricity bills. Demand was generally too low, and revenue did not increase sufficiently to cover the increased expenses. He therefore stopped using his electric sewing machine. Some women explained to us that the tailor spent all the revenue to pay for the electricity for the machine.

This basic challenge of starting a business that would create more revenue than costs, including the cost of electricity, was also a part of other people's considerations regarding the potential use of electricity for income generation in this village. During the open-ended interviews, six people mentioned the idea of buying a freezer in order to make ice and ice-cream as a business. They knew that such businesses existed in two of the other villages, Maka Sarr and Ndombil, and sometimes they bought ice or ice-cream from these villages.

The number of households in Sine Moussa Abdou with such business ideas was likely much larger than the number of such businesses that would saturate the market for ice and ice-cream in the village. When we mentioned this concern, one man told us that he would be able to succeed with a freezer business because he was already a businessman and he and his family knew how they could run a business with a freezer. However, he also added that currently, power is expensive. The business that his family was already running was to trade fruits and vegetables between villages. They had a horse, and the man traveled from village to village to trade. He said that if they had money, they would start a business that the women in the family could engage in.

One woman in Sine Moussa Abdou told us that she had had a freezer previously and made ice-cream and ice for sale. Her monthly revenue had been between 5,000 and 10,000 XOF (between 7.6 and 15.2€), but the power bill was too high to make this business profitable. She explained that the freezer required too much power because it was old and too big for the business she ran. Enersa had recently started to test and give advice regarding energy-efficient freezers and provide special connections for freezers because they require a 24-hour power supply.

In Maka Sarr, one woman who ran a freezer business selling ice and cold water paid 16,800 XOF (25.6€) per month for the freezer connection. She said that business was not always good but that she did not lose money. The monthly surplus after paying for the electricity was about 15,000 XOF (22.9€).

The challenges involved in generating income through use of electricity mentioned by these respondents are likely to be typical. The possibilities depend on how much revenue increases due to using electricity and whether this electricity use provides savings on energy expenses. For instance, it could be possible to achieve savings when somebody has been using diesel generators before and then switches to solar power instead.

Affordability and other economic considerations on the part of users

Generally, the affordability of the electricity services was relatively good in the studied mini-grids. We state this based on the high portion of households connected, as described above, and the number of appliances people were using, i.e., the amount of electricity they could afford to use. The number of households in the studied villages that went beyond the lowest price option also supported this point. According to the survey, 78% were using Blocks 2, 3, and 4. However, two-thirds of these households were using Block 2.

Complaints about the price level despite money saved for many customers

Social inequalities clearly made electricity, even the cheapest Block, more difficult to afford for some. Some of those who were connected (especially to Block 1 but also to the larger Blocks) struggled to pay their electricity bills because of the fluctuating incomes and the challenge of collecting a larger amount at a time as compared with a previous bit-by-bit payments for candles, torch batteries, and kerosene. According to our survey, it was common to pay too late and thereby have to remain without electricity for a short period. A majority of the survey respondents (77%) said that there were times where they failed to re-charge their cards due to an inability to pay. "If we don't have money, we stop paying and use torch and candles," one of them said. Short periods of a few days without electricity seemed to be most common, according to people's descriptions in the open-ended interviews. This was supported by the company, which reported that their records showed few variations in payment per year. Two of the customers told us that they had stayed without electricity for 2–5 months due to an inability to pay. Most likely, based on these various observations, some customers had difficulties in affording electricity. The electricity company allowed some flexibility, accepting that people would take breaks in payment, even though its plan had been to collect the payments without interruption. The company did not want to force people to pay and felt that such individuals would not be able to pay in any case. In some cases, relatives and family members working in urban areas assisted with electricity bills, even sending money for electricity payments directly to the local operator or seller.

When people did not pay on time, one hypothetical reason could have been a lack of interest in having electricity continuously. However, the interviews showed that this was not the case, as explained above. The customers expressed a strong interest in using electricity, both through words and through their actual use patterns. Enersa also reported that people truly wanted to have the power because they began calling their office after 30 minutes if there was a power outage.

Another observation regarding affordability was that a portion of the population was likely to save money on the use of electricity (and receive much higher-quality energy provision) as compared with their previous energy expenditures (on candles, kerosene, torch batteries, radio batteries, phone charging, etc.). As mentioned in Chapter 3, the average expenditure per month was 12,829.05 XOF (19.5€) according to a survey carried out in Sine Moussa Abdou by Peracod and Inensus in 2009. We do not know how many people spent less than the average, but many likely reduced their spending, depending on the type of Block. The price for a Block 1 supply was 5.7€ (3,739 XOF) per month, as well as an 8–14% additional installment payment per month for some of the internal wiring, as explained in Chapter 4. Block 2 cost double this amount.

Subscriptions for the two cheapest options, Blocks 1 and 2, taken together, accounted for 67.5% of the initial connections, with the latter accounting for nearly half (or 45.5%) of the total connections. Second, slightly over a third of respondents (32.5%) subscribed the most expensive blocks, Blocks 3 and 4, with the latter being limited to 6.5% of initial subscriptions. The third observation relates to the dynamics of subscription switching over time. At the time of the survey, 6% of respondents had downgraded from Block 3, while 2.5% had downgraded from Block 4. Conversely, subscriptions to Block 1 had grown by 6%, and those to Block 2 had grown by 2.5% at the time of survey, relative to the initial choices. The field interviews indicated that downgrading from a higher- to a lower-cost block was one of the coping strategies used by some households facing financial difficulties caused by a drop in income from farming, casual labor, and/or remittances.

Despite the point made above that most basic electricity services were affordable for the majority of the populations in the four villages, the respondents (except for one household) universally claimed that electricity was too expensive. This complaint about expensive electricity was also reported by those who subscribed to Blocks 3 and 4. In the qualitative interviews, people normally brought up this issue without prompting.

We found that there was a combination of reasons for this complaint. In addition to issues of affordability due to the irregular and low incomes of most of the people, another important reason was that the people compared their electricity prices with those in grid-connected areas, where the price per kWh was lower. People knew the details of the Senelec supply through friends and relatives living in grid-connected areas, including cities such as Dakar and Thies, and through visits to grid-connected areas. The fact that the wealthiest complained as well and that the arguments used were similar across the population indicated that this was an established discourse in the villages, partly as a matter of principle. It is likely that the political promises regarding lower and uniform tariffs were known to them. The respondents thought that the monthly expenses would have been less if the national grid had been available in their villages. On this point, the seller in Ndombil commented that he did not agree and argued that if the Senelec grid had come there, people would not have been able to pay for all the electricity they wanted to use.

Another reason for the complaints was that people aspired to use more electricity than they were able to afford. People also compared themselves with others who were able to use more electricity than they were; for instance, those who watched TV in their homes. Many people were keen to watch TV, which required at least 2–3 Blocks, and some wished to use freezers, fridges, fans, computers, and other appliances.

Factors that enhanced affordability

In addition to analyzing the reasons for people's complaints about the tariffs, it is also important to consider how the Enersa business model achieved relatively good affordability for the village citizens despite having a higher price per unit of electricity than the national grid. One important reason for the fact that nearly everybody in each village population could afford to subscribe was that the project implementer attempted to assist people in extracting as much energy service out of each unit of electricity as possible.

The following features of the system design contributed to this: The subscriptions were based on the types of energy services available and offered some very low-level options that provided the most basic energy services (indoor and outdoor lights, phone charging, and radio) for a limited number of hours per week. Moreover, the costs for the connection fee and the electrical wiring in the buildings could be paid in small installments over a long-term period. One feature of the design that was meant to help people in budgeting for electricity but did not work as expected was the smart card, which could have helped customers to check how much electricity was left based on their prepayments. Furthermore, an emphasis on energy-efficient lights (CFL) helped users to extract much more light per kWh than incandescent lights. The latter were cheaper to buy, so they were still used by some, but this reduced the benefits they derived from the electricity supply. One problem for some was that their TVs were old and consumed too much electricity.

Apart from these features of the socio-technical design of the electricity system, the local, socio-cultural context and the socioeconomic conditions in each geographical area strongly affected affordability. Compared with villages in different counties in Kenya in which we had performed research previously, people in the Senegalese villages studied here could afford much more light and other services. In Kitui County in Kenya, for instance, people struggled to afford one light at 0.09€/¢ (10 Kenyan shilling (KES)) per day, equivalent to 57 XOF per day. In the Enersa mini-grids, the cheapest option provided several lights and phone charging at about 0.2€ (132 XOF) per day on average. Thus, in the case of Enersa, the Senegalese people could afford to spend more than people in the Kenyan villages, and they also received more electricity services for the money than the lantern renting provided in some places in Kenya. The sharing of a connection in each large extended family living in the same compound also enhanced affordability.

Affordability was also enhanced by accepting that the customers took small breaks in electric bill payment and the use of electricity when they had economic problems. This helped them to deal with fluctuating incomes. Moreover, the grant funding for parts of the investment cost was also an important factor because it reduced the portion of those investment costs that would have to be recovered through the sale of electricity.

One observation regarding how the tariff level for electricity provision from mini-grids is judged in the general discussions of this topic is that it is normally compared with grid electricity on the basis of electricity units (kWh typically), not on the basis of how it fulfills people's need for energy services. For solar home systems (SHS), the opposite is true, with the focus being on the price for energy services. If the price of the energy services received through SHS were measured per unit of electricity, this price would be very high.

According to this research, the project implementers could do more in this direction, for instance, by helping people to gain access to other kinds of energy-efficient appliances and by providing additional information about how they could get as much out of the electricity they buy as possible. One example would be to inform people about opportunities to use "extra energy," as explained below.

Views on the 28-day payment schedule

Another common complaint in relation to payment was about the tariff system. As described in Chapter 3, the common payment system practiced on the company's side was a four-week payment schedule. The customers paid for 28 days at a time, while the supply limits per customer were programmed for weekly periods. The seller of the electricity entered the prepaid period into the electronic card of the customer before the card was inserted into the special meter.

Some people found it frustrating that the payment did not cover an entire month but only 28 days. If they paid on the first of a certain month, they could not pay on the same date the next month but would have to pay 2–3 days before. This was perceived as somehow unreasonable or illogical. "We wish it could be 30 days or a month because 28 days is like a month minus 2–3 days," one man explained, "and we wish to have it cheaper." One woman in Léona told us,

> We pay once per month. We go to buy power, and the man who sells it tells us it is for a month. Then, we don't know whether it is for 28 days or 30 days or 31 days. Sometimes, the card is empty before the end of the month. For instance, if we buy on the 1st in a month, the card is empty on the 25th. Then, we go directly to the operator and buy more. We get frustrated because the month is shorter.

In this case, the frustration is both because of the 28-day payment schedule and because the respondents use up the power before the payment period expires. "I know it is really hard for the head of our household to pay the bill, and I wonder why, after buying for a month, it doesn't last for a month," she added, happy to be able to express her views on this issue because she had been thinking about it for a long time.

In Maka Sarr and Ndombil, respondents complained about "the 28 days" from a different perspective. People said that the payment expired after 28 days, even if there was money left. One man said, for instance, "we never finish the power before this." Thus, they felt they were using less electricity than what they had paid for. Another old man argued, "this is something that really matters. The power is programmed for 28 days. You never know whether there is money left on the card or not. They just cut it off." People wished to be able to know their balance to better manage their consumption. They found it unfair that the payment expired even when they had used very little electricity, for instance, when they had been away for a few weeks. People in Maka Sarr said, "if we don't use it, we lose it." This problem was not mentioned in Sine Moussa Abdou, probably because that village had an older system and people there had developed higher levels of consumption. In the newer systems, people tended to use less than agreed-upon and wanted to pay less.

Both types of customers (both those who used all of their power before the end of the prepaid period and those who felt there was power left when the card expired) needed to know their balance in order to adjust their consumption. In order to check the balance, the card had to be read in a customer's meter, but the customers did not keep the cards. The company had intended to let the customers have the cards all the time, but some complications occurred in this regard. The company found that people were not careful enough with the cards. Some people broke or lost them, as Enersa staff reported to us. Thus, the sellers kept the cards, but they could not read the balance on the cards when the customers came to pay for the next month. To do so, the sellers would have had to walk to each customer's house. Our Inensus contact person told us that he saw a need for mobile phone solutions for payment in order to resolve these issues and that Inensus was considering this in new systems in other countries. Inensus had developed the meters without mobile payment because such payment is not yet widespread in Senegal, while it has become common in East African countries. This is an example of how the broader socio-cultural context influences the direction of socio-technical innovation.

Some people requested that the company simply meter their consumption of electricity so that they pay for their actual consumption, just like people do in places where they have a connection to Senelec. Inensus reported that its customers had knowledge of the details regarding the power supply from Senelec. Inensus also reported that it is working towards payment on a kWh basis for higher consumptive load, such as for productive enterprises within their new projects.

When Inensus developed its business model and implemented it through Enersa, it found the Block system to be important in planning installed solar PV capacity. It also expressed that if people do not use the capacity, this represents a loss for the company. They saw an arrangement in which people "use it or lose it" as logical in relation to solar PV technology because the electricity from each solar panel (which can be seen as parallel to a Block) is generated whether it is consumed or not. The idea was therefore that the customers should pay a fixed amount every month. This would also be similar to the grid in that those connected to the grid also have to pay a fixed amount per month, as well as for actual consumption. Thus, the supplier found this to be a fair way of doing business with the customers. "Sell it or lose it" could be a way to describe this thinking. However, the customers wanted to pay for what they consumed, not what was produced at the power plant from each solar panel.

A smart way of using "extra energy"

It was possible for the customers to purchase additional electricity when they wanted to, as explained in Chapter 4. This could be used both to extend the supply period and to increase the load. This "extra energy" had the advantage of allowing any load size, in contrast with the load limits of the Block system. The "extra energy" was more expensive per kWh than the regular electricity, but it could be purchased for as little as 1,000 XOF (1.5€) at the time or in any other amount. Furthermore, this extra energy would not expire and could make an important difference for the customers. For instance, 1 kWh purchased for 1,000 XOF (1.5€) would make it possible to watch TV while subscribed to Block 1 because the power limit would no longer be applied for this energy. It could also help alleviate the fact that the prepaid time periods expired too early.

How fully this opportunity was used varied by village. According to one person at the company, it was used most in Sine Moussa Abdou, Maka Sarr, and a third village not covered in this study (Merina Diop, commissioned in September 2014). As of March 2016, 1 year and 2 months after start-up, Ndombil has had only one customer purchase extra energy according to a company representative. In Sine Moussa Abdou, the operator (seller of electricity) informed us that six customers had purchased extra power in March 2016. All of these had subscribed to Block 1. He also mentioned that a family with Block 2 had purchased extra energy when they had a festival. The seller also used this option in his home. He had a Block 1 sub-scription, and by purchasing extra energy for 1,000 XOF (1.5€) every month, in addition to the 4,400 XOF (6.7€) that he paid for Block 1 per month, it became possible for him use a black-and-white TV for short hours. This was cheaper than having Block 2. He had probably explained this more economical opportunity to the other people in the village as well.

Quality of electricity services and user satisfaction

Various characteristics, or attributes, of the electricity supply determine how useful the electricity services people can obtain are. ESMAP (2015) mentions attributes such as availability, reliability, capacity, convenience, and affordability, which was analyzed above. Usability is crucial as well, as is analyzing the reliability of the supply (stable voltage, few and short power outages). Because usability is so important, we consider opportunities to use electricity for various purposes (loads and timing possibilities), as well as the suitability and practicability of the electricity supply in view of people's needs for electricity services. We also include the brightness of the light bulbs allowed within the system, which affects the benefits derived from having access to electric light. Another attribute that we see as important is the users' freedom to increase or vary their demand, in other words, their flexibility in terms of how they can use electricity. This makes it possible to vary the level of expenses they incur and thereby improve affordability, because most people's incomes fluctuate over the year in rural areas.

The reliability of the power supply includes the frequency and duration of power outages, as well as the stability of the electric current in terms of voltage. The latter was not mentioned as a problem by any of the people interviewed in the villages in Senegal. Regarding power outages, there were two main types. One type was power outages related to the power plants and distribution system. The other type was power outages within an individual connection. The first type was not affected by the users' actions, while the other type was affected by such actions.

Power interruptions in the mini-grids took place to a varying extent in the four villages. We do not have precise data about these interruptions, but they were not highlighted as a serious problem by the customers. In one of the villages, Léona, there was a technical problem for the time being because one of the two inverters had been sent to Germany for repair. This was necessary due to warranty conditions, because this type of inverter (Sunny Boy) did not have any authorized repair facilities in Senegal or the neighboring countries. Thus, there were more interruptions in the supply than usual. In Ndombil, people reported power outages lasting for a few hours that occurred several times per week. One reason for the interruptions was increasing demand and the need to increase the capacity, which was being planned. According to the Inensus contact person, the battery was cutting out more often because of fast-growing demand and limited capacity, and the diesel generator was running more often in this area.

Short internal power outages within the individual connections were very common. One reason for such power outages was the overloading of appliances by the customers. If they exceeded the power limit for the Block they had subscribed to, the smart meter would automatically switch the electricity supply off for this connection. The electricity would come back on automatically. Some customers told us that it would come back on after 30 minutes. Others reported that the power did not come back on automatically, but it was easy to confuse this with the energy-related power outages explained below, as there was no indicator showing when an overload had occurred. Power outages caused by overload were a built-in part of the system, and their number depended on the users' own ability and willingness to stay within the power limits of the kind of connection they had subscribed to. Whether they managed this depended on how well they understood these regulations, as well as their willingness and ability to limit the use of power.

Other reasons for the electricity supply being interrupted within individual connections included that the electricity the customer had paid for was used up, or that the period for which the payment was valid had expired, as explained above. This seemed to be a challenge for the customers, although this was given less attention by them in the interviews than the issues of "use it or lose it" and the 28 days. It could be difficult for customers to stay within the agreed-upon amount of electricity they could use in a week. If they had consumed more than the agreed-upon 1.4 kW per Block, the power would go off towards the end of the week. This would happen even though people had paid for a longer period (normally 4 weeks). This was because the customers had exceeded the units of electricity programmed for consumption per week in their meter. Examples of this were especially common in Sine Moussa Abdou, where the power supply had operated for the longest period (6 years). The Enersa staff confirmed that the weekly programming was working as the customers described.

The limitations regarding electricity consumption were similar to those of a solar home system, because the units of electricity available in a given period were limited. The difference was that in a solar home system, the user would budget the electricity consumption per day, knowing approximately how many hours certain appliances could be used. Here, budgeting the power is done for a week at a time, providing more flexibility in terms of varying electricity consumption but also making it more difficult to have an overview of usage.

We have argued above that one important aspect of the quality of electricity services is flexibility (Ahlborg 2015; Ulsrud et al. 2015). The Enersa mini-grids were characterized by high flexibility for the users, much more so than other solar micro-and mini-grids. There was flexibility in terms of the various subscriptions, and flexibility in terms of the timing of consumption. Electricity could be used at any time, and the amount consumed could be varied from day to day. The possibility of receiving additional power and thus adding larger loads also provided flexibility. This flexibility was facilitated by technical and organizational solutions, such as the smart meters, which facilitated the Block system, the upfront payments, the company's willingness to accept breaks in supply, and the use of a diesel generator. The generator could be switched on when there was high demand for power. The degree of flexibility was further influenced by the power capacity of each power plant relative to the local demand for electricity.

However, as mentioned above, user satisfaction was not only affected by the functioning of their power supply but also by comparisons people made with other solutions. In this case, comparison with electricity provision by the national utility Senelec was especially prominent, which was natural given that the Senelec grid was very near. Some said that there were no limits on how much electricity could be used there. A woman in Ndombil, for instance, said, "I envy people in a village called Ngakham because they have Senelec power. When I am there, I watch TV from morning to night" (laughing).

People's way of understanding the electricity system

The design of the mini-grid systems appeared to be slightly too complex for some of the users to understand it easily, even given the thorough information provided by Enersa. This played a role in how the system worked for the users and their satisfaction with it. When asked about the reasons for the internal power interruptions, some people said that these were caused because they had switched on too many appliances at once, while others said that they did not know, and some suggested that the problem was at the power plant. For instance, members of a household in Sine Moussa Abdou showed an understanding of the power limits (the possible loads) by explaining that in order to be able to watch TV on Block 2, they switch off all their lamps. They also showed understanding of the energy limits (the amount of electricity that could be used over time) by saying that they use the TV for around 2 hours during the night and the lights for around 4 hours during the evening and night. Some women said they had learned to switch off some lamps if the power goes off. They found that it is possible to use four lamps at the same time as the TV. A woman in Sine Moussa Abdou told us that she did not know why the power ended 2 days before the week was over, and we asked her the following question: "Have you asked somebody why it happens?" Her answer was, "I have asked the seller, and he asked me if we were using too much electricity. I told him no, we are not using too much." Q: "Do you know for how much time per day you can use electricity?" A: "We have no idea of how many hours we can use it per day."

We also had the following conversation with some smiling women inside a relatively wealthy home with a Block 4 connection in Léona village. This is an example of people experimenting with how much power they can use by plugging in appliances that need more power than what the subscription allows. An iron draws much more power than the 200 W limit of Block 4, and this appliance seems to be of interest in all the areas in which we have performed our case studies. An iron normally requires about 1,000 W of power, but we have been told in Kenya that "solar irons" now exist, i.e., irons that require much less power. The conversation went as follows:

Q: "Do you ever notice that the power goes off because you are using many appliances?"
A: "Sometimes, when using an extension cord, where we plug in the TV, if we want to use the iron in the same extension cord, it doesn't work. The iron works if we put it directly into the plug. We tried to plug the iron and the TV into the extension cord, but it burned (laughing)."

These examples show that there were aspects of the system that some of these informants did not understand in the way the project implementer had intended, although the work of information spreading had been taken seriously. However, importantly, the examples also showed that most people had learned about a range of aspects of the system, both through the numerous information measures enacted by Enersa and through practical experience.

Gender sensitivity appeared to be a relevant theme in this regard. The information provided by the company mostly reached the men because they would represent the family in meetings and because the men would be seen as the natural contact person during initial household visits. Women would be likely to be present in the home, but if a man were also around, he would be the one who communicated with the representative of the electricity provider. How much information this individual household member passed on to other members of the household likely varied by household. Women's knowledge about the electricity system seemed to be based on learning by using it. The following conversation with a woman provides a further illustration of this:

Q: "Are the power outages in your house affected by anything you are doing?"
A: "No, there is nothing we are using that would provoke such outages. We only use lamps and mobile phones, while in other houses, people are connecting many other apparatuses, which can lead to power outages in their houses."
Q: "Are there other reasons for the outages? Has the seller explained them to you?"
A: "No, the seller just talks to the men, not to the women. I only know from what I see in my house."

The most difficult issue to understand is the energy limit (the limit on energy consumption over time), which has two variables: load and duration. It is simple in theory – if a consumer switches on all the appliances allowed within a given block (the entire possible load), they can be used for around 4 hours per day. However, in practice, it is difficult to understand how a certain increase or reduction in load affects the allowed duration. According to our open-ended interviews, some people seemed to understand this, while others did not. An example of the latter was an old couple in Sine Moussa Abdou. We asked them, "Do you think people here in Sine Moussa Abdou understand that the amount of electricity depends on both the amount of power you use at a time and on the number of hours this power is used?" The question was put like this to avoid focusing on their own knowledge. They answered, "No, nobody told us this." They also said, "Before the end of the week, we don't have power for one or two days, and by the beginning of the next week, it comes back." Q: "Every week?" A: "Yes, it goes like this."

Even for those people who fully understood the system, it could be a challenge to manage and budget electricity use. Due to the existence of large families, a large number of people switched on the lights or charged phones in different rooms of a given house or in different houses within a compound, and all these various electricity users may not naturally coordinate their use of power. This is relevant regarding both load and the duration of the electricity supply. According to qualitative interviews conducted in the various villages, family members tended to switch on the lights and TV whenever they wanted. "We use electricity according to our needs," one woman said, for instance. Some men blamed women for wasting electricity. One man told us,

We men are, in a way, obliged to buy a connection because our wives and children want electricity like other people, so we, in a way, have to connect, but as time goes by, it becomes difficult to pay for the electricity, so we may just stop using it.

The seller in Ndombil stated that he had begun to talk with women and children also, in addition to men, to teach them how to save power.

We asked the seller in Sine Moussa Abdou the following question: "Even if people understand the rules, is it difficult for them to limit their use of electricity? Is this because the planning and monitoring of a family's electricity use over a week is likely to be a complex task, especially given large families living in several buildings or rooms and many people using electricity?" He answered, "Yes, of course. Some people who come to me, if they have not reached their 28 days, say 'I know it is because I overuse power.'" Furthermore, the seller in Ndombil told us that it was easier for people when smaller family units had separate connections, for instance, each brother and his wives having a single connection, instead of the old father, his adult sons, and their wives all sharing a single connection.

The opportunity for increased consumption, new connections, and access to alternatives

In some mini- or micro-grid systems, it can be difficult to continue connecting new customers after demand has reached the limit of the installed capacity. Expansion may be hindered by a lack of economic surplus in the operation of the system and by the lack of a good strategy or mechanism via which to gradually expand along with increasing demand (Ulsrud et al. 2011).

One of the innovative elements of the socio-technical design studied in Senegal was the annual reviewing of people's contracts. People could decide to purchase more or less capacity and sign a new contract. The company would then have documentation of the need for capacity increase and be able to add equipment based on this. However, this mechanism had not yet become very important. In the pilot village (Sine Moussa Abdou), the initial capacity was higher than needed, and after 6 years, it had still not been exhausted, although demand had increased gradually over the years. During the last few years, there had been a 10–15% increase in electricity consumption per year according to Enersa, but the new customers and increasing demand from existing customers could still be easily included in the system. In three of the other villages, where the electricity and supply had operated for from 1 year (Maka Sarr) to 1 year and 9 months

(Léona and Wakhal Diam), the limits of the installed capacity had been reached in only a short time. One power plant already required expansion (Ndombil, after 1 year and 3 months). According to the project implementer, demand was growing quickly in Ndombil, so the diesel generator at the power plant was running much more often than in the other villages. There were also power interruptions because of the limited capacity as compared with the demand. The plan was to expand the power plant soon. It was therefore likely that new customers could continue to be connected continually in all the six villages.

Apart from expanding the capacity of the power plants, the expansion of the mini-grid networks was also sometimes needed. In Sine Moussa Abdou, for example, a new household was going to be connected soon, and several new poles were installed. According to one of the company staff members, this was necessary even though the house was very close to the existing settlement. The financing of such an expansion was a challenge, according to the project implementers, because it would normally require additional support, which is difficult to obtain. GIZ and other funders had provided grants for the initial grids. In particular cases, the company itself covered the cost.

Alternative solutions to obtain access to electricity, especially for lighting and phone charging, were available and used both by the customers, as a backup and complement, and by people without a connection to the mini-grids. The majority preferred to use the mini-grids as their main source of electricity supply, and in addition, most of the customers used other sources of electricity and lighting (including non-electrical sources). Out of the 46 power plant customers interviewed in the survey, 13% reported using solar home systems. A family in Maka Sarr, for instance, already had a solar home system when they connected to the mini-grid, and they continued to use this system for watching TV and sometimes also for light and phone charging. They had used the solar home system for 30 years. Regarding lighting devices, a majority (85%) of those interviewed revealed that they used torches as an alternative source of lighting, in addition to the mini-grid. They also used candles (34%), phone flashlights (23%), solar torches (15%), solar lanterns (7%), and kerosene (2%), as shown in Figure 6.2.

The battery-powered torch, solar torch, and phone flashlight were used daily by many power plant customers, while candles and kerosene were occasionally used.

A relatively new type of lighting source that has seldom been mentioned in the literature on off-grid lighting is the use of flashlights on mobile phones. Phones commonly have a built-in torch that can be switched on when needed. Because people without a power supply can charge their phones in the neighborhood, the use of mobile phones to receive a small

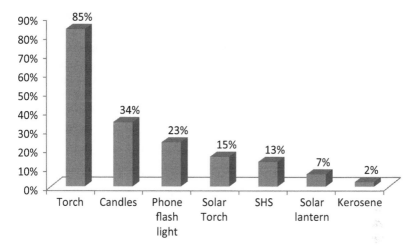

Figure 6.2 Alternative sources of lighting used by mini-grid customers

amount of light is a practical option. This option may provide better light than candles and hurricane lamps, depending on the type of phone. One woman in Sine Moussa Abdou who could not afford to pay the electricity tariff told us that her family used mobile phones for lighting daily and that the charge lasted for 4 days at a time. Eighty-three percent of those who used phones as flashlight did so daily.

The few respondents who were not connected to the grid reported that electricity had benefits for the village, including phone charging at neighbors' premises, street lighting, and businesses for community members. Additionally, those who were not connected appreciated the fact that they were able to buy ice-cream and ice from grid-connected shops and watch television at their neighbors' houses.

The role of geographical distance from the power supply

Geographical distance from the power plant to people's homes is often a problem in relation to mini-grids because of technical features that limit their outreach and because settlements are often more dispersed than the outreach of the grids. Also, for conventional grids, geographical outreach is a challenge because of the costs involved in expanding it to all parts of dispersed settlements and because of voltage drops when lines are long and demand is low.

In this case, geographical distance was not a problem, due to dense settlement patterns, which were very well suited to mini-grids. The villages were built as coherent units, and the distances from one side of each village to the opposite side were short – no more than a 10-minute walking distance. The radius of the Sine Moussa Abdou village, for instance, could be estimated at less than 1 kilometer. In dispersed types of settlement patterns, complementary solutions would be needed, in addition to mini- or micro-grid-based electricity supplies.

Summary

Based on qualitative and quantitative data from Sine Moussa Abdou, Maka Sarr, Ndombil, and Léona and interviews with the various people involved in the mini-grid projects, this chapter has analyzed a range of characteristics of the electricity access provided by the Enersa mini-grids in these villages, according to dimension 5 (The Electricity Services) of our analytical framework. The chapter has also added to Chapter 5 by showing additional aspects of how the mini-grid systems functioned in practice and the factors that had an impact on their operational and economic sustainability (as part of dimension 4, Functionality).

The chapter has shown that affordability was relatively good but that due to the deep poverty of significant portions of the population in these villages, the majority chose one of the two cheapest options or was still sometimes not able to pay for electricity. Those who could afford the more expensive options had more freedom to use electricity, but they still complained about high prices. The reasons for people's complaints about prices were both comparisons with the national utility, Senelec, and their own economic struggles due to irregular and low incomes. As is typical in rural villages, there were large internal variations among the families, and therefore, some would save money by using electricity instead of the candles, batteries, and kerosene they had used before, while others would not.

We have also shown that people's economic problems became mixed with frustrations about the payment system and ways of potentially controlling and managing electricity consumption. At the same time, our data leave no doubt that people nevertheless strongly appreciated the benefits of having good light, easy phone charging, and other electricity services and saw this as a significant and positive change in their lives. The power Block system contributed to better affordability due to differentiated connection fees, and affordability was further enhanced because the project implementer aimed to provide the customers with as much energy service out of the electricity supply as possible in terms of services such as lights, TVs, fridges, etc.

People were interested in using more electricity and had ideas about how to generate income via the use of electricity. This chapter has explained why it was difficult for them to achieve this, although there were a few positive examples. The mini-grid systems were well suited to the socio-cultural context in which they were located. Their design made it possible for almost everyone to use light in most of the rooms and outdoor spaces in which they performed their daily practices at home.

The conditions of power consumption appeared to confuse some customers, despite the company's willingness to provide information. Maintaining an overview of connected load multiplied by hours of use over the course of a week in a large family with many rooms and buildings is not an easy task. However, the weekly allocation of electricity provided some flexibility to the customers, for instance, as compared with a solar home system, in which power consumption is directly affected by daily charging and the size of the system.

Dissatisfaction can potentially lead to consequences for the functioning of the system due to unwillingness to pay and the unregulated use of electricity, such as use of much stronger light bulbs, charging batteries for others, and illegal connections (Ulsrud et al. 2011). In previous cases, this has affected the operational and economic sustainability of the systems by reducing the lifespan of the batteries and reducing power plant revenue despite increased consumption of electricity. In the case of Enersa, however, such consequences were prevented by the power and electricity limits of the Block system, including prepayment. This was also positive in terms of user satisfaction because it prevented a downward trend in service quality.

The reliability of the electricity supply was good in general, but automatic power outages in the homes were common. These outages were not due to technical problems but to a control mechanism for avoiding electricity loads that were higher than planned for and for stopping the supply when the paid amount was used up or expired. These mechanisms were triggered relatively often due to people's desire to exceed the power limits and problems with the budgeting of electricity use over a week's time. The experienced reliability level of the power supply was therefore lower than the performance of the power plants and mini-grids themselves. This chapter has demonstrated the value of analyzing the dynamics between the energy system and the socio-cultural and socioeconomic characteristics of the society in which it is placed. The chapter has also illustrated the importance of a qualitative understanding of electricity access that is supplemented with data revealing the quantitative aspects of that access.

Note

1 The number of households covered was 52.

References

Ahlborg, H. (2015). Walking along the lines of power: A systems approach to understanding co-emergence of society, technology and nature in processes of rural electrification. *PhD thesis*. Göteborg, Sweden: Chalmers University of Technology, Department of Energy and Environment.

Cabraal, A., Cosgrove-Davies, M. & Schaffer, L. (1996). *Best practices for photovoltaic household electrification: Lessons from experiences in selected countries.* Washington, DC: World Bank.

ESMAP. (2015). Beyond connections: Energy access redefined. ESMAP/Sustainable Energy 4 All (SE4ALL). Technical Report 008/15.

Palit, D. & Chaurey, A. (2011). Off-grid rural electrification experiences from South Asia: Status and best practices. *Energy for Sustainable Development*, 15 (3): 266–276.

Ulsrud, K. (2015). Village-level solar power in practice: Transfer of socio-technical innovations between India and Kenya. *PhD dissertation*. Department of Sociology and Human Geography, Faculty of Social Sciences, University of Oslo.

Ulsrud, K., Winther, T., Palit, D. & Rohracher, H. (2015). Village-level solar power in Africa: Accelerating access to electricity services through a socio-technical design in Kenya. *Energy Research and Social Science*, 5: 34–44.

Ulsrud, K., Winther, T., Palit, D., Rohracher, H. & Sandgren, J. (2011). The Solar Transitions research on solar mini-grids in India: Learning from local cases of innovative socio-technical systems. *Energy for Sustainable Development*, 15 (3): 293–303.

Winther, T., Ulsrud, K. & Saini, A. (2018). Solar powered electricity access: Implications for women's empowerment in rural Kenya. *Energy Research and Social Science*, 44: 61–74.

7 Replication – influenced by factors at multiple scales

The sixth and last dimension of our framework, on replication of the decentralized systems for electricity supply, is analyzed in this chapter. For a thorough explanation of how we define replication, see Chapter 1, dimension 6. As an example of replication, and to highlight further lessons learned by Inensus, we here introduce Inensus' next step. Near the end of their struggle in Senegal, Inensus decided to start a new mini-grid project in Tanzania, creating a joint venture called "Jumeme." Their first mini-grid in Tanzania was inaugurated in April 2016, immediately after our fieldwork in Senegal. Inensus brought many elements of their model to the new venture, but they also made several changes and adaptations as per the new socio-cultural, socioeconomic, political, and geographical context.

Therefore, when discussing the replicability of the Enersa mini-grid model below, we sometimes refer as an illustration to Inensus' own replication of the model in a different context. Jumeme is located near Mwanza in Tanzania, on Ukara Island, and the plan, according to Inensus, is to spread it to other islands in the Tanzanian part of Lake Victoria. We have not visited the place to collect data, so the information we have is from conversations with Inensus. If research funding were available, it would be useful to perform a similar case study on the Tanzanian case and to monitor it over time.

Replication of a given electricity model, as we show below, depends on most dimensions we have analyzed here. It is often assumed, even by those who plan a project, that the replicability mainly depends on the characteristics of the model itself. However, it also highly depends on the contextual conditions, including how influential actors view the model. Various actors' judgments on the way the model works are guided by their values, their willingness to financially support the model, or their attitudes towards profitability and the private sector's role. Replication also depends on the space for maneuvering of the type of actors implementing it, be it small or large companies, public sector units or donor agencies, as well as national regulations and international trends (Ulsrud 2015). An important part of the context is the wider socio-technical systems of the energy sector, both conventional energy regimes and emerging systems (or niche systems), as described in Chapter 1 as part of our analytical framework. Examples of such contextual factors are the availability of suitable financing or suitable technical equipment in a specific country. The socio-cultural and socioeconomic features of the geographical area (described as part of our analytical framework) also influence the opportunities for replication, in both similar and different places. We discuss these aspects of replication below in terms of how they affected both the replicability of the Enersa mini-grid model in Senegal and the Jumeme work in Tanzania.

Features of the national and international context that affected replicability

Inensus' work in Senegal ended because of a policy that was never implemented in practice, as explained in Chapter 2. The ERIL program, with its promising invitation to private sector companies to contribute to off-grid electrification in Senegal, only resulted in *one* ERIL license and concession contract for a private-sector-led mini-grid. This was the one given to the pilot project in Sine Moussa Abdou after 4 years of effort by Inensus and Enersa, while the other five, that had been constructed in parallel with this struggle, ended up operating without a license. Chapter 2 explained that due to the government's decision to establish a system for harmonized electricity tariffs, the government did not accept the tariff level required by Enersa to make the business viable because it was higher than SENELEC's. In our interviews, government officials mentioned ideas for filling this kind of tariff gap through cross-subsidies and still expressed some interest in private sector initiatives on mini-grids. However, the work on this had not come far and was moving slowly at the time of writing this book. As part of their struggle, Inensus and Enersa tried to change the directions taken by policymakers toward future policies in this field, but according to their experience, there was nobody in the government who really understood what was required to have the private sector involved.

Replication had been just about to become a reality in the Enersa mini-grid activity in Senegal when it had to be canceled because of regulatory and political factors. Both Inensus and Enersa had put significant efforts into the planning of 24 additional mini-grids. Financing and grant funding was in place, and the sites had been selected through a systematic process of gathering and analyzing information. The political and regulatory factors that led to lack of licenses, and thereby investor withdrawal, blocked this work for Enersa and in general made it difficult or impossible for private sector companies to start mini-grids in Senegal.

A feature of the donor community that affected replication was their lack of effort, and inability to assist a government that was in the middle of comprehensive, long-term, and demanding reforms where ambitious policies pushed through by the World Bank and other powerful actors were supposed to be implemented. This entailed a lot of risk and uncertainty for the government and a need for other kinds of or more expertise than they had.

Although this kind of private-sector-led mini-grid model clearly could not be replicated in Senegal, a similar model might be replicable in other countries with different policies and regulations. Such replication could be accomplished by either Inensus or other distinct mini-grid implementers (not merely private sector companies). Moreover, some elements of the model could be more replicable than others.

Inensus started their new activity in Tanzania because they identified better opportunities than in Senegal. The Tanzanian regulatory framework and policies were quite different from the Senegalese and seemed to be more suitable for a private-sector-led business model, such as Inensus', although there are uncertainties regarding the political support for future mini-grids. Enersa has continued to operate their six mini-grids in Senegal up to the current, but they struggle economically and have lost some key staff. The future of the company Enersa is quite uncertain.

Features of the socio-cultural context that affected replicability

In addition to the policy framework and other factors at the national and international levels, the features of the different local socio-cultural contexts where the mini-grid systems are located also play a role in what kinds of models can be replicable and in which places. In Chapter 6, we argued that the Enersa model was well adapted to the social realities on the ground in several ways. Here, we briefly discuss how the relations between an electricity system and its local context might affect replication. In the Enersa case, it was demonstrated that many people could share an electricity connection, such as due to the organization of their extended families. This made it possible for Enersa to electrify 1,000 people through fewer than 70 connections. Although this probably saved costs, it also caused Inensus' sophisticated system for allocation of power and electricity to work less smoothly than they had expected, as explained in Chapter 6. The project implementers learned that it was better to split up the connections more so that smaller family units could subscribe in the following villages. There will certainly be different socio-cultural factors that are relevant in different places, and these must be studied and taken into account by project implementers.

As is broadly known, settlement patterns and terrains are important for the opportunities for provision of electricity through mini-grids. The settlement patterns in Sine Moussa Abdou, Ndombil, Maka Sarr, and the other villages served by the Enersa mini-grids were dense and compact and therefore very suitable for mini-grids, although mini-grids can also be used where the settlement is slightly more dispersed. For a general discussion on this theme, see for example Palit and Chaurey (2011).

The size of the villages, the poverty level, and existing economic activities also play a role in replication. A lesson learned by Inensus and Enersa is that they must select larger villages with more economic activity to achieve a better economic performance, meaning that the Enersa model might be difficult to replicate in small villages unless the operations are subsidized. One of the problems is the facilitation and initiation of income generation for the population by the use of electricity, as analyzed in Chapter 6. As expressed by Inensus, new income-generating activities were very difficult to achieve in these Senegalese villages. According to Inensus, a mini-grid business cannot survive purely by providing household services; it is also necessary to have larger customers, which are difficult to find in small villages. At the same time, increased income generation and electricity services are desperately needed also in small villages.

Features of the mini-grid model and the project implementers that affected replicability

Those characteristics of the mini-grid model itself that are most likely to be important for the opportunities to replicate it in large numbers are the investment costs, economic design and performance, organizational and operational setup, and time and resources required for the implementation process and follow-up (Ulsrud 2015). Moreover, different types of actors have different opportunities to replicate a given model.

In the model studied here, the investment costs for distribution and generation were separated from each other and financed in different ways, as explained in Chapter 4. The financing for the fixed assets, including the electricity grid (or distribution network), came from grants and potentially from public sources. The financing for the power generation was commercial, based on the aim of creating a commercially profitable/viable activity. As a result, the investment for the distribution system would rely on grants from others, while generation would depend on the company's ability to raise funds for a commercial activity, recover the investment, and earn a surplus.

Replication of the investment would thereby depend on the willingness of funding organizations to give grants to or in other ways support this kind of electricity model, as well as on the economic performance of the power supply and the company in general. GIZ offered grants to Inensus' pilot projects. The interest of the German government in supporting the use of solar PV technology in the South has been strong for many years and seems to be continuing. The German government supports activities on mini-grids in Nigeria, Kenya, India, and Tanzania. The EU has also supported some of the same projects, including Jumeme. A Dutch fund supported the five solar mini-grids implemented after the pilot in Senegal. It is not clear whether such funding of mini-grid distribution networks is available for any company or organization that has carried out a promising pilot project. We do not have insights into who can obtain such funding from GIZ or other agencies. Some governments provide relevant support in their countries.

Replication of the commercial part of the financing of this kind of model depends on the conditions for commercial loans. Such conditions vary according to the policy of each bank or financing institution involved, such as their interest in social goals, including electricity access for all and business with a social motivation, and their willingness to set long time frames. Such patient funding is required due to project risks related to cash flow and security. The conditions for commercial loans also vary by country. Financers who have purely commercial interests would be likely to find better business opportunities elsewhere than in the kinds of rural areas described herein, where people's incomes are among the lowest and their ability to pay for electricity is very low. It is precisely these areas where people lack access to electricity and mini-grid companies such as Inensus and Enersa hope to create a win-win situation between meeting social needs and simultaneously achieving a profitable business.

The long-term economic performance in operation and maintenance of the mini-grid systems also has a strong impact on their replicability. This includes, in addition to covering the salary expenses and other running costs, the chance to pay back the loans and provide the returns expected by the investors. Investors want to see quick results (according to the Inensus experience), and project implementers might realize that they have promised their investors too much in terms of the economic performance and how long it takes to reach profitability. We do not know exactly how the Enersa mini-grid model was performing economically or whether the potential 30 mini-grids that were planned could have been a profitable undertaking. However, according to the project implementers' calculations based on their pilot experience, profitability would have been possible. Their choice to commit to the venture indicates that they had good reasons to believe that it could work as a business. There would still be uncertainties, however, due to the risk of unexpected problems.

A challenge experienced by Inensus through both Enersa and Jumeme, which also has implications for replicability, is that the growth in electricity sales has been slower and at lower levels than hoped, for the reasons explained in Chapter 6. However, the opportunities are better in the Tanzanian project, especially because the project implementer selected larger villages with more ongoing economic activities. The plan is to implement 20–30 power plants during the next couple of years, with a target of 10,000 connections.

In Tanzania, where the first step is a mini-grid on Ukara Island, the project implementers also put additional effort into facilitation of income generation for the citizens, seeing this as a key to making the mini-grids economically viable. They cooperate with an organization that focuses on this and is present over time. This work is funded by the project's budget. The activities include ice making (for chilling of water, air, etc.) and chicken incubators. Water pumping is provided with the excess electricity, and enabling a rice irrigation project. However, it still takes time to increase the sales of electricity. Moreover, replication of this element of Inensus' work would require similar grants and arrangements with organizations working on income generation.

One factor that makes replicability more difficult is that mini-grids cannot yet match the tariffs common in utility supply (which are usually regulated and not cost-reflective). This renders the mini-grid companies dependent on convincing governments and customers of the necessity of having higher tariffs in mini-grids. The issue of cost-reflective tariffs versus harmonization of tariffs, as well as subsidies, is further discussed in Chapter 8.

Replication of the organizational, institutional, and operational setup of the Enersa mini-grid model would most likely be possible in other places. It functioned well in many ways and did not seem to be dependent on factors that would be difficult to find elsewhere. For instance, identifying people who could take on the tasks performed by the sellers/operators or the village electricity committee would be likely elsewhere as well, although this would vary. According to Inensus, it is necessary to have skilled people available in the villages who can be trained as operators, village electricians, or members of a village electricity committee, and this could be difficult in some places. Similarly, the role of a village electrician would require relevant background knowledge. One feature of the model that enhanced the chances to replicate it was that it was not too dependent on the performance of such individuals due to the use of company staff for many tasks.

A part of the organizational aspects, the Block system could be replicable, but Inensus changed this and developed the model further when taking it to Tanzania to start mini-grids there through Jumeme, adapting to the new social context. The customers in Tanzania were not in favor of the Block system and wanted metered electricity. It was easier for the project implementer to allow for metering of electricity consumption here due to the larger size of the village and concomitant larger potential for demand growth. Moreover, instead of using smart cards, Inensus (through Jumeme) is working with the phone network operator Vodafone to start using mobile money for the tariff payment. Another difference regarding the tariffs is that in Tanzania, a 50 W Block would have been far too much for the lowest type of subscription, because only one light (8 W CFL) would be suitable for some customers. There is also a kind of pay-as-you-go tariff for those with the least ability to pay, who use very small quantities of electricity. The unit cost for this subscription is high, but these customers can pay and consume when they want to.

Another element of replicability is the time and effort required by the project implementers for the implementation process and the operation, maintenance, and follow-up for each village's power supply. This strongly influences the operation and overhead costs in terms of time, office, and travel costs for the project initiators, such as Inensus. In Senegal, the project implementers (Inensus and their Senegalese joint venture Enersa) invested much time in the pilot project in the first village. Starting a company, planning, financing, and carrying out a project represents a complex process, according to the project implementers. In each of the five following Enersa mini-grids, the process was much less time consuming.

When it comes to the normal follow-up of operations over time, the project implementer in this case expressed the importance of limiting the time spent in the villages, since this is the highest expense related to the operation. However, good follow-up is important to performance. Our Inensus contact told us that implementing the power plants in clusters is therefore essential for the chances to implement larger numbers of mini-grids.

The concept of long-term, society-wide transitions (see Chapter 1) is useful as a reminder that the introduction of new socio-technical solutions, such as mini-grids, does not happen in isolation from broader social processes and is not limited to developing the details of the specific models in themselves. Moreover, these so-called socio-technical experiments might be important drivers of larger changes, especially if they expand to larger and more significant activities. The literature on transitions and the construction of new, comprehensive systems for fulfilling energy needs or other societal needs show that a range of other changes are required beyond the development of scalable business models. There is an interaction between the individual projects and the changes in wider socio-technical systems of energy sectors and also between such energy system changes and processes in the wider societal contexts outside the energy sectors, including politics and international power relations. The opportunities of the individual actors, such as mini-grid implementers, depend on the interests of powerful actors, some of whom are working against change. Moreover, some of those who are promoting change, even toward similar goals, sometimes choose different strategies than would be suitable for small actors, such as Inensus. This is a power struggle, often implicit, and a discursive struggle, where the flagship reports, speeches, buzzwords, and big numbers tend to dominate, while the more practical discussions on what works and how it can be made to work better are less visible. An example of this is the way hundreds of other mini-grids are now being implemented in Senegal, as mentioned in Chapter 2, following approaches that have not been functioning well in the past.

Summary

Since replication (or upscaling) of the Enersa mini-grid model had to be abandoned in Senegal, the superficial conclusion may be that the model is not replicable. However, it is possible to implement a model similar to the Enersa mini-grids in larger numbers in a more conducive policy environment than the Senegalese. Whether it would actually be effective is an empirical question that cannot be answered without real-world application, but in our analysis above, we explained why it is plausible. The answer would partly depend on what kind of economic performance is expected and how much economic support is provided to the projects. Moreover, *elements* of the model can certainly be repeated in other cases.

Without doubt, the Enersa experience has contributed strongly to the learning on how renewable and hybrid mini-grids can be designed, implemented, operated, maintained, and replicated, and this knowledge can be used in new mini-grid activities. The new solutions being tested currently, by both Inensus and other mini-grid developers, seem as if they can solve some of the problems that confronted Enersa, but the actual performance of the new projects cannot yet be known. In the following chapter, we propose general conclusions from this case study, based on our thorough analysis of the Enersa mini-grid experience.

References

Palit, D. & Chaurey, A. (2011). Off-grid rural electrification experiences from South Asia: Status and best practices. *Energy for Sustainable Development*, 15 (3): 266–276.

Ulsrud, K. (2015). Village-level solar power in practice: Transfer of socio-technical innovations between India and Kenya. *PhD Dissertation*. Department of Sociology and Human Geography, Faculty of Social Sciences, University of Oslo.

Ulsrud, K., Winther, T., Palit, D. & Rohracher, H. (2015). Village-level solar power in Africa: Accelerating access to electricity services through a socio-technical design in Kenya. *Energy Research and Social Science*, 5: 34–44.

8 Conclusions, part one: Lessons for how to do mini-grids

The point of departure for this book was that decentralized models for electricity provision, based on solar and other renewable energy sources, are necessary for the achievement of sustainability goal number seven on sustainable energy for all, and that mini-grids are expected to be one of the main decentralized models for the future, but they still meet many hindrances. The book has aimed to contribute to important kinds of knowledge for the achievement of such goals and visions for an equitable and sustainable global society.

This analysis has therefore been guided by the overall solution-oriented research question of how to achieve well-functioning, equitable, and usable mini-grids that can be implemented in significant numbers. The research on such village-scale electricity systems was motivated by the large potential for decentralized solar power in the future, and we analyzed an example that we found to be broadly relevant. We have followed the journey of the entrepreneurs and drivers of a small-scale solar-hybrid mini-grid activity in Senegal, Inensus, and their Senegalese joint venture Enersa. The analysis of this journey has been based on a framework of analysis that directed attention to broad, interacting dimensions of a certain type of energy model, solar mini-grids.

The most striking characteristic of the mini-grid case analyzed here is the contrast between the positive practical achievements on the ground and the political, regulatory, and financial barriers that blocked further upscaling. One might therefore ask whether this case was too special to be relevant for mini-grids in general. Despite the problems that thwarted the upscaling, however, we find that the achievements and lessons learned on practical, technical, and organizational aspects of the mini-grid model are highly relevant. Electricity access was provided for a large portion of the population in each village served by Enersa, and the operation and maintenance mostly functioned well. Moreover, the kinds of barriers met at the national-level are also relevant for many countries in Africa, Asia, and elsewhere, and the case represents a useful starting point for discussing how such problems should be solved.

In the summaries of the chapters above, we have provided conclusions on the characteristics and performance of the Enersa mini-grids and the factors influencing these outcomes. In the following, we conclude on the overall lessons that can be drawn from this case study and consider their relevance for other projects, in Senegal and in other countries. In the two first sections, we present conclusions about factors that influence the functioning of small-scale renewable-energy-based mini-grid systems, especially those with solar PV as their main generation technology, in terms of both operational and economic sustainability. Thereafter, we present conclusions about factors that influence the quality of the electricity access, and finally, we briefly present conclusions about factors that influence replicability, but this topic is further treated in Chapter 9.

These three aspects correspond with three of the six dimensions of our framework of analysis and have been analyzed in Chapters 5 to 7. When discussing these below, we bring in the other three dimensions wherever relevant: The role of national and global framework conditions, including political factors (Chapter 2), the role of the local context (Chapter 3), and the role of the project implementer's visions, socio-technical design, and process of implementation (Chapter 4).

Factors that influence the operational functioning

The Enersa case shows promising achievements for the operational sustainability of future mini-grids. We defined operational sustainability as the electricity system's ability to keep operating in practice and continue to reliably deliver its electricity services without significant reductions in the quality of supply (see Chapter 5). Typical challenges for the operational sustainability of mini-grids have been poor maintenance, lack of motivation among operators, technical weaknesses, lack of suitable metering devices and control mechanisms, and insufficient training and support (Camblong et al. 2009; Ulsrud et al. 2011). Political factors at the local level have also played a role in some cases. Some of the practical challenges of small-scale renewable mini-grids have been solved over the last 5–10 years, and Inensus has also contributed to this. Their Enersa project shows that it is possible to make these mini-grids function well in several ways, although there is still potential for further improvement, as Inensus is addressing in their recent activities.

The most important aspects that influence operational sustainability include the functioning of the technical equipment, the procedures needed to run the electricity system, and the arrangements for management, prompt decision-making, monitoring, maintenance, and expansion. In addition, the interaction with the customers for delivery of electricity services and payment for these is also crucial.

Even though this case study demonstrates how deeply the social dimensions influence technological change, and although technical experts have increasingly recognized that a number of nontechnical aspects are important, the technical design is still critical. Technical design and installation have to be performed through proper estimation of demand and careful engineering. The devices and systems for metering electricity consumption and preventing unregulated use are central, in addition to good controllers and inverters. As shown by the Enersa case, such devices and systems prevent typical problems observed in previous mini-grid systems, such as over-consumption, lack of payment, and fast degradation of the batteries.

There seems to be a need for more technical innovation in this domain, based on familiarity with the socio-cultural and physical context for use and project experiences. All technical installations have to be adapted to the conditions in rural areas, such as large amounts of fine dust from dry soil. The technical equipment not only must have advanced functions but also should be easy to use, robust, and durable. Equipment design must take into account that not everyone will handle equipment carefully and correctly.

Project implementers often purchase from a different part of the world to get robust, good-quality equipment. A disadvantage is that repair might become complicated and time consuming if certain equipment has to be shipped far away. This could create long periods of power outages, as in one of the mini-grids observed in Senegal, but it might be a necessary yet rare disadvantage of choosing high-quality equipment. For the future, it might become possible to develop good maintenance facilities in a larger number of countries, including regional-level maintenance facilities covering neighboring countries of a particular region. This would be a part of the gradual development of novel socio-technical systems in the energy sector that facilitate the use of emerging technologies.

Operational sustainability also depends on suitable routines for operation and maintenance of the electricity system, accomplished by people with the right skills. One central choice is how to distribute the various tasks between technically skilled outsiders and people who live in the villages. Village inhabitants are capable of taking care of many tasks, with the right training and follow-up, while outsiders might still better handle tasks that require additional technical and managerial expertise. Enersa seemed to have found a good balance, as shown in Chapter 5. However, even after ensuring that the tasks are suitable, the people to be trained sometimes need to have certain skills from the outset.

In several mini-grids, the performance and commitment of the individuals working as operators has been decisive to the functioning of the mini-grids, and these individuals' tasks have often been difficult due to technical shortcomings (Ulsrud et al. 2011). Even with better equipment, the performance of individuals matters. To depend less on the performance of individuals, a relevant strategy is to hire people with more education and experience who can travel, visit the local operators regularly, carry out most of the maintenance and repair of technical equipment, and simultaneously follow up with and encourage the local operators. This method functioned well for Enersa, supporting findings from other cases of village-level electricity supply (Ahlborg and Sjöstedt 2015; Ulsrud et al. 2018).

The main challenge with frequent follow-up is that it requires time and money from the project implementer. Locating the mini-grids in clusters is one way of reducing expenses for time and transport. Another way is to reduce the follow-up, but this choice has to be weighed against the importance of making the operation, maintenance, and customer management function well. There are mini-grid systems that minimize company visits in the communities served, through robust equipment, remote monitoring, and centralized prepayment. However, a risk is that if project owners are rarely present in the communities after implementation, necessary learning processes on the local situations and needs might be lost, regarding the views and practices of existing customers and the perspectives of those who consider connecting but require more information.

This study also suggests that the social engagement, values, and inspiration of the project implementers might be important for the achievement of well-functioning mini-grids or other technology projects in areas with a high poverty level. Enersa's achievements were due to committed efforts, as we have also seen in other cases. Project implementers with purely commercial interests would probably not invest in such areas. A risk is that the project implementers' commitment might decrease over time if the problems they confront are not solved despite significant long-term efforts. One example is when other actors, such as the government or donor partners, are unwilling or unable to do their part.

The Enersa case nevertheless supports common assumptions on the private sector's advantages in creating innovative, well-functioning, efficient operation of infrastructure. Private-sector-led implementation and operation of mini-grids seem to create a strong incentive for ensuring well-functioning systems in the long run, because companies that operate their own mini-grids have a profound interest in maintaining sufficient functioning over time. They have invested their own money, which they want to recover over time through successful operation and sale of electricity.

Factors that influence economic sustainability and profitability of mini-grids

Economic sustainability and profitability is clearly more difficult to achieve than operational sustainability (Ahlborg and Sjöstedt 2015; Ulsrud et al. 2015). In addition, in the Enersa case, despite the strong profitability goal, the economic performance was weaker than the project implementers expected. Economic sustainability is the electricity system's ability to cover its own costs and for small, village-scale mini-grids this might not be realistic from a short-term perspective. If it can be achieved, however, it eases the process for project implementers, governments, and donors. Subsidy programs are an option, but creating an efficient, long-term, and broadly available system to subsidize operation costs for renewable mini-grids owned by the private sector would be a complex task because of variations between the business models used and the need for new governmental procedures that could be perfectly reliable for the companies. Therefore, we find economic sustainability to be a desirable characteristic of private-sector-led mini-grids.

As suggested in many reports and research publications on rural electrification, an important reason for low economic performance is the *low ability to pay for electricity* among the people who live in the areas receiving access to electricity for the first time (Ulsrud et al. 2015; Palit et al. 2011). In the Enersa case, there is also little doubt that people would have used much more electricity if they had been able to afford it. This was clear from people's frustrations about the power limits of their subscriptions, their attempts to plug in more appliances than subscribed for, and the fact that they used up their pre-purchased electricity faster than the intended time. Most complaints were related to the wish to use more electricity, including comparisons with the utility electricity supply and objections to the tariff levels.

There are better chances to achieve good economic performance in larger communities than the ones served by Enersa, not only because of better chances to achieve income generation for the users of electricity but also because the villages are likely to have a higher number of people with ability to pay for electricity above the very basic level of a few lights and mobile charging. In the new projects in Tanzania, Inensus focuses on much larger communities, as mentioned. However, the small communities also need the kinds of electricity services they can provide, so it is still relevant and important to implement mini-grids in small villages, such as the ones studied in Senegal. The challenge is to create realistic economic arrangements for this.

Another oft-mentioned factor is that "anchor customers" can support the economic performance of the mini-grids. This might succeed in some cases but also creates some vulnerability, because one or a few large customers determine the mini-grid's economic performance; if they close down, this can cause a breakdown of performance.

Due to such constraints, people's consumption of electricity is likely to increase slowly, but the increasing trend can continue for a long time. In the Sine Moussa Abdou village, for instance, consumption was still increasing every year after more than 6 years of operation.

A strategy to achieve both good economic performance and affordability often suggested in discussions on electricity access for all is to *facilitate income generation* through the use of electricity. This increases the demand for electricity and thereby the economic performance of the electricity system while simultaneously reducing poverty for the people in the villages. This can potentially create a virtuous circle of economic improvement for both the company and the population. However, income generation as a key to better economic performance of off-grid electricity systems appears to be difficult to achieve in practice. There is a lack of results of this kind, and this research has contributed to understanding why.

The best opportunities for profitable income generation via electricity exist where electricity can provide a cheaper and more practical option to replace something else, as with the mill observed in the Ndombil village (see Chapter 6). The most relevant example is people who are already using diesel generators for businesses. The business can make more profit by switching to electricity, because the electricity produced by solar and hybrid mini-grids is likely to be cheaper. There may also be supply chain issues to procuring diesel in remote rural areas. Switching to cheaper electricity leads to increased income generation for the owner, perhaps more employment, and even, eventually, more demand for electricity from those who have increased their incomes and thereby can afford to pay more.

However, if the ambition of the electricity provider is *new* income generation, beyond replacing existing and more expensive energy solutions, it seems to be necessary to facilitate income generation through targeted initiatives as part of the implementation of electricity provision. People do not have enough information about all the potential opportunities of electricity and might not be aware of which machines exist and which kind of products/services can be produced and sold, and they might need training and financing support.

As shown in Chapter 6, new income generation by the use of electricity is difficult in small, remote communities. The challenge for the citizens is to earn enough on the new activities to have a surplus after paying the electric bill and other expenses. However, even in the small communities served by Enersa in Senegal, there were minor examples of "productive use" of electricity, while others had tried and thereafter given up. A general challenge in this field is to identify other actors who can work with the mini-grid companies as advisors and facilitators of income generation by the use of electricity, not least of which is to find or establish markets for products from marginalized areas.

In addition to ways of increasing revenue, *innovative cost reductions* are also necessary to achieve economically sustainable mini-grids. There is a dilemma between cost reduction and creating jobs through mini-grid implementation, including jobs with the electricity supplier itself. The pressure to reduce operation costs leads to limited employment and low salary levels for local operators, while jobs and better salaries are desperately needed in such places. The pressure to reduce costs might also lead to less facilitation of income generation with the use of electricity. Furthermore, when limiting the size of the power plant to reduce the maintenance costs, especially in terms of the size of the battery bank and thus the replacement cost, less capacity will be available for income generation and innovative ways of using electricity in the communities.

In general, achieving economic sustainability presents the challenge of balancing different desirable characteristics of mini-grids, such as the amount of electricity provided (i.e., the access tier achieved), the costs of installation and operation (especially the capital costs but also the costs of battery replacement and other operation and maintenance), and affordability. Increasing the access tier level means increased costs but not necessarily higher incomes for the implementer, since people are not always able to benefit from the higher tiers. A high overhead cost is also a challenge for mini-grid companies, putting a heavy burden on the business. The chances to reduce costs have improved to some extent due to price reductions for the equipment, especially the solar PV panels.

An important factor influencing economic sustainability is the cost and lifetime of batteries. As shown in several previous studies on off-grid electricity supply, problems have been found related to batteries, not least the replacement cost. In this Senegalese case, such problems have not yet occurred, as explained in Chapter 5. Nevertheless, battery replacement is part of the calculation of the operation and maintenance costs, thereby also influencing tariffs. If battery costs were to decrease, this would help mini-grid developers to reduce tariffs and come closer to matching the main grid's tariffs.

There is a dilemma between rural electrification in general and the economic performance of the electricity system. This is the case both for extension of the main grid and for mini-grids and other off-grid models. Balancing affordability and economic performance is a key challenge for all electricity provision in rural areas, where poverty is large and costs for providing electricity are high. Finding smart ways of fulfilling these two qualities of electricity provision is at the core of the ongoing efforts of private sector mini-grid developers.

Lessons on how to provide good-quality electricity access from mini-grids

The quality of electricity access includes affordability, accessibility, suitability/ usability, and reliability. These affect user satisfaction and people's opportunities to use and benefit from electricity services.

Comparisons with the perceived functioning of other solutions affect users' satisfaction

A factor that may contribute to complaints about tariffs and other dissatisfaction is comparison with and perception of the functioning of other alternatives, such as the national grid. Even if the tariff for a mini-grid connection is higher than the national electricity tariff, it might be expected that people who lack access to grid electricity would be willing to pay for this and be grateful for alternative solutions. If the alternative solutions are more expensive per unit of electricity than the national grid is, people might still find this acceptable, if the grid were not an option. A further assumption would be that people would compare the mini-grid services with the energy sources they used before, which had poorer quality and cost more per month for many. Also compared with solar home systems (SHS) and smaller solar PV systems (pico-solar), the price per kWh of electricity from mini-grids is usually lower.

However, even though people highly appreciate the upgrade from what they used before, they might tend to compare with what others have that looks more appealing. Even if people save money on the power supply, they might still be dissatisfied because they find it unfair that people in other parts of the country pay less.

People in areas where the main electricity grid is not present also tend to assume that the electricity supply from the main grid is functioning well, which is often not the case. Poor reliability is a major problem in grid supply in a range of countries. In a recent research project in Kenya, for instance, we noted that the grid supply was so unreliable in two study areas (one in eastern and one in western Kenya) that it cannot be counted even as a tier one level according to the UN's multi-tier framework (Winther et al. 2018).

If both the mini-grid and main grid systems function well, the main grid is still likely to present better opportunities. But there are also similarities. A prepaid subscription will cut off power when it has expired for both a main grid and a mini-grid customer. The load available would be limited in the cheapest and most affordable subscriptions for both the main grid and a mini-grid system, and a mini-grid might offer some cheap options for subscription that go below the cheapest options from many national utilities.

Comprehensive information increases benefits and improves affordability

According to our research, comprehensive information to those who might connect is important even when designing a system that is easy and simple to use. Information and communication is also important after implementation, when people have the chance to experience the infrastructure in practice and better know what questions they would like to ask. People who develop a good understanding of the system can benefit more despite the constraints of affordability. This is also beneficial for the project implementer, because the customers will be more satisfied and consume more electricity. For instance, with sufficient understanding of the system, people might learn how they can choose subscriptions with the best fit for their own needs and economy. People can also benefit more from flexible arrangements, such as the "extra power" offered by Enersa. This option was not often applied by the customers, but those who had tried it realized how it could improve their own access to electricity and used it to sometimes go beyond their load limits and energy limits (the number of kWh they subscribed to per week or month).

Such solutions that appear useful for some of the customers in one village could be described for leaders and customers in other villages to facilitate the same there. This could help the customers to optimize their benefits despite the strong limits to what they can afford. This would increase the users' satisfaction, and it could in turn benefit the project implementer by increasing the sales of electricity.

A gender-sensitive approach enhances the usefulness of electricity access

A gender-sensitive approach is likely to be important in electrification (Winther 2008), as illustrated by the Enersa case, which shows that not only addressing the genders equally but even addressing women *more* than men is likely to be positive for all parties involved. Those who communicate with existing customers and potential future customers should address women at least as much as men. It is still common to think that electricity is a men's issue, even in homes, but women are often the ones who deal with it the most on an everyday basis.

In the case studied here, the women had a large impact on how electricity was used, since they stayed at home more than men did. They needed light for their household work in the evenings and early mornings, and they appreciated watching TV as a source of information and entertainment for themselves and their children. The women found out how to handle the electricity provision at home despite little direct contact between them and the project implementer. If they had been offered direct information and the chance to ask questions, this could have made them better able to utilize the opportunities provided by the system and understand the constraints. They would then be able to increase the benefits for themselves and their families. A constraint for reaching the women during information campaigns was probably that Enersa mainly had male employees, so men were the ones who visited the villages. When they came to a house, the men living in the house tended to appear. If a woman had visited, it might have been easier to address the women. In general, involvement of both women and men in the supply of electricity would be likely to enhance the benefits for both genders.

Affordability is not only about the tariff per kWh

Something that is little acknowledged by customers and observers of mini-grids is that some project implementers try to facilitate as much energy service from each unit of electricity as possible. In the Enersa case, one example was advice on efficient refrigerators. Another example is packaging energy services (such as light and TV) in blocks. Although an important motivation for this was to help Enersa in the planning of the capacity of the power plant, it also allowed people to get a sense of what kind of electricity consumption was possible for them. However, there is potential to do more. Energy-efficient appliances (low-power irons, for example) can help the customers to get more and better energy services out of each unit of electricity and thereby improve the user satisfaction and affordability of certain electricity services, such as watching TV. More efficient TVs would probably have made an important difference in the Enersa model. Many people had old TVs, and some customers on the two lowest levels of subscription owned a TV but did not have enough power to use it. This seemed to be one reason for complaints about the tariffs. TV was important for people, since it gave them a range of new inputs and insights, according to their descriptions of what watching TV meant to them. The fact that some people were able to watch TV at home, when the majority could not, seemed to create some frustration for people in the latter group.

The case study shows that affordability of electricity provision is affected not only by the electricity tariff in terms of price per unit of electricity but also by how the energy services are provided. This includes access to energy-efficient appliances and how the subscription is designed (for instance, broadly differentiated connection fees and tariffs and different caps on consumption). In this case, the two cheapest blocks were especially important to reach the majority of the population. These provided relatively low levels of service, but they still offered more electricity than the common types of SHS.

Such broad access to electricity services, as in the villages served by Enersa even if only very basic for the majority, is still a distant goal for most electrification efforts, both off- and on-grid. The socioeconomic conditions (i.e., people's income levels) also played a role in this achievement, as the income level in the studied villages might be higher than in several other areas, but the design of the subscriptions was nevertheless very important.

Despite the higher tariff than that of the national utility for customers connected to the main grid, the majority of citizens in each village served by the Enersa mini-grids were able to afford an electricity consumption that was between tier one and tier two of the UN's multi-tier framework. This shows that the design of the tariff system and service delivery compensated to some extent for the electricity price. Additional energy-efficient appliances would also help.

Inensus' ongoing mini-grid work in Tanzania has taken such ideas further, and the practical experiences with this should be studied.

Factors that influence the replicability of an electricity model

Some key questions about replication are whether it is possible to repeat the same arrangements in large numbers and how the necessary organizational solutions can be developed and adapted to different kinds of societies. Replication of small, decentralized mini-grids is a challenge, but there are many current ongoing initiatives, according to information from policymakers and energy experts. The difficulty however, is not so much to plan and implement them, but most of all to ensure operational and economic sustainability in a large number of rural villages.

The Enersa example in Senegal provides some insights in replication of private-sector-led projects despite the situation that forced an end to it in this case. The issue of how to design regulations for such mini-grids is currently a challenge for policy makers, and we discuss this and other issues related to replication in Chapter 9.

References

Ahlborg, H. & Sjöstedt, M. (2015). Small-scale hydropower in Africa: Socio-technical designs for renewable energy in Tanzanian villages. *Energy Research and Social Science*, 5: 20–33.

Camblong, H., Sarr, J., Niang, A. T., Curea, O., Alzola, J. A., Sylla, E. H. & Santos, M. (2009). Micro-grids project, Part 1: Analysis of rural electrification with high content of renewable energy sources in Senegal. *Renewable Energy*, 34 (10): 2141–2150.

Palit, D. & Chaurey, A. (2011). Off-grid rural electrification experiences from South Asia: Status and best practices. *Energy for Sustainable Development*, 15 (3): 266–276.

Ulsrud, K. (2015). Village-level solar power in practice: Transfer of socio-technical innovations between India and Kenya. *PhD dissertation*. Department of Sociology and Human Geography, Faculty of Social Sciences, University of Oslo.

Ulsrud, K., Rohracher, H., Winther, T., Muchunku, C. & Palit, D. (2018). Pathways to electricity for all: What makes village-scale solar power successful? *Energy Research and Social Science*, 44: 32–40.

Ulsrud, K., Winther, T., Palit, D. & Rohracher, H. (2015). Village-level solar power in Africa: Accelerating access to electricity services through a socio-technical design in Kenya. *Energy Research and Social Science*, 5: 34–44.

Ulsrud, K., Winther, T., Palit, D., Rohracher, H. & Sandgren, J. (2011). The Solar Transitions research on solar mini-grids in India: Learning from local cases of innovative socio-technical systems. *Energy for Sustainable Development*, 15 (3): 293–303.

Winther, T. (2008). *The impact of electricity: Development, desires and dilemmas*. Berghahn Books, Oxford, UK.

Winther, T., Ulsrud, K. & Saini, A. (2018). Solar powered electricity access: Implications for women's empowerment in rural Kenya. *Energy Research and Social Science*, 44: 61–74.

9 Conclusions, part two: The structural challenges

Important lessons on solar and hybrid mini-grids and other renewable-energy-based, private-sector-led mini-grids have been presented in Chapter 8 above. In this final chapter, we would like to reflect and conclude on some more overarching issues relevant for how mini-grids can become both successful and widespread. Decentralized electricity systems like solar and hybrid mini-grids make it possible for a diversity of actors to engage in electricity provision. These actors are relatively free to create innovative socio-technical designs within the constraints of costs, existing technology, people's needs and challenges, and other factors mentioned in Chapter 8. However, despite this freedom to innovate and the flexibility in how to design such village-scale electricity systems, the success of these energy systems are also highly dependent on the broader energy system context at national and global levels. The established energy systems (or regimes) are so far not well equipped to deal with these small newcomers in fruitful ways, and changes in the existing institutional structures are therefore necessary. The opportunities to achieve such changes, as seen in Senegal, are likely to be influenced by path dependency in the energy sector in terms of established ways of doing things (like the model for tariff setting created 20 years ago by the World Bank) and economic and political interests in the broader political economy.

We first provide a discussion of what should be the future role of small-scale renewable mini-grid systems. Thereafter we discuss potential solutions to issues about tariff setting, which are significant challenges for the future of mini-grids. We point out choices and required efforts for policymakers, donors, and investors, if they want to include small-scale, renewable energy mini-grids in the future energy mix. In the final section, we reflect upon the usefulness of the framework of analysis and theoretical concepts used.

The future role of mini-grids

Given the complexities of mini-grids, would it be better to concentrate on other models for electricity provision? Without mini-grids and other village-scale models (e.g., energy centers and charging stations), two main options would remain for areas lacking or with dysfunctional electricity provision: extension of conventional electricity grids (including large, conventional mini-grids) and stand-alone systems for individual households, businesses, and public facilities. While important and necessary, these two models have some shortcomings when it comes to providing electricity for all. The shortcomings of the conventional grids to provide universal, reliable, and affordable supply are well known (see Chapter 1). This is why off-grid solutions, both village-scale models and stand-alone systems, have become a significant and rapidly improving alternative. The stand-alone systems are mainly solar lanterns and SHS of varying sizes/capacities. Stand-alone systems make a difference for many people. So far, they are the most widespread of the off-grid electricity access models. Yet, stand-alone models also have their shortcomings. For instance, the types of systems that dominate the market are small or pico-solar systems (e.g., less than 20 W). These provide basic services only, such as lighting and cell phone charging, and even these are only affordable for certain parts of the population in each rural community (Bloomberg 2016).

Village-scale mini-grids, where the settlement pattern is sufficiently dense, can complement stand-alone systems to both include a larger portion of a rural community in the electricity supply and utilize the potential of solar electricity for a much wider range of services, including the use of appliances that demand a high-power capacity. Inensus' Senegalese mini-grid model analyzed in this book demonstrated both of these features. Such provision of electricity services also frees the users from the responsibility to invest in equipment, battery replacement, and other maintenance. For these reasons, mini-grids are relevant, even though they also have shortcomings.

So is it necessary to have private-sector-led provision of electricity, which is usually a public service? Private sector companies, both small start-ups and large companies, are important drivers for the social and technical innovation of these models. Governments can also implement and operate small-scale mini-grids, but private sector companies have demonstrated that they are nevertheless important for the progress in this field. Without the private sector's efforts to develop new electricity models that work in poor, remote communities, a rich source of innovation on such models is lost, as is clearly illustrated by this case study. Such innovation in turn is important for government-led implementation of mini-grids.

Furthermore, private-sector-led mini-grids are important because the task of providing electricity access for all requires a wide range of technology options and investments as well as operational and management models pioneered, tested, and facilitated by different actors. It should be noted that a private sector actor was in the Enersa case the main driver and responsible actor for both implementation and long-term functionality. This is different from a more common situation where companies are just suppliers to the government, either of technical equipment and installation, or of operation and maintenance services.

In order to survive and grow, a private sector mini-grid needs a revenue from the tariffs that the customers pay for the electricity. The tariff should be affordable for people at the same time as it covers the expenses for the supply of electricity. In addition, it should ideally not be higher than the tariff for electricity from the main grid. This book has shown the difficulties of balancing these ideal goals, and below we discuss potential ways forward, underlying political and economic interests and potential institutional work.

Should mini-grid tariffs reflect the full costs of providing the electricity?

The main policy question about tariffs is whether they should be harmonized across different electricity models and diverse geographic regions and populations. A pressure to harmonize tariffs is likely to emerge in many countries when regulations for mini-grids come up for actual consideration, as we have observed in Kenya. This is likely to happen when the request for licenses and pressure to work out the details of the regulations pushes the authorities toward making decisions. When or if a government flags that it accepts more expensive electricity in remote areas than in the grid-covered areas, it is likely to meet resistance. It remains to be seen what different governments decide on this matter.

One implication of harmonization of tariffs would be some kind of support or cost compensation, such as cross-subsidization for electricity suppliers that operate in the most difficult areas, until they succeed in matching the general tariff through cost reductions and higher revenues. This inequity is acknowledged by offering cross-subsidization, as conceived in Senegal, but it seems to be a large and complicated task to define it, implement it in practice, and ensure that it is a reliable support mechanism for private sector companies or others who operate mini-grids or off-grid electricity systems. In some countries, such as Kenya, there is a well-functioning regulated system for cross-subsidization within rural electrification carried out by the government, but only internally within the government.

As part of such a strategy, it would be useful to inform the public of how the conventional tariffs are calculated and how much these are supported by direct and indirect subsidies. This would enhance the understanding of why the mini-grids also need subsidies.

Private sector mini-grid companies, however, have strong concerns about a dependence on subsidy payments with varying levels of structure and reliability from governmental units. Such dependence would create substantial vulnerability for their operation as cost and capital recovery becomes uncertain.

Another aspect of harmonization of tariffs and related subsidies is that if the government (as part of the provision of infrastructure and welfare) wishes to harmonize the tariff, then it also has the responsibility for bearing the cost of providing the subsidy. Then new questions arise, such as whether the government has the financial resources for this additional cost and how these resources can be raised, such as through cross-subsidies. Cross-subsidies might be a politically sensitive issue where it is not already a normalized governmental strategy. Typical problems of subsidization also must be addressed, such as distorting market activities where the private sector is already delivering some related products and services.

If the governments do not have steady financial flows to provide subsidies to companies or government units operating in areas where it is more expensive to deliver electricity, they should probably avoid harmonizing the tariff. A harmonization in such cases would make it impossible for small mini-grid companies, such as Inensus, to contribute to electricity provision, as we have seen in Senegal. This would then harm and delay the innovation and ongoing cost reduction for solar mini-grids and other models suitable in many geographical areas in developing countries. Thus, governments must take many pros and cons into account and debate these thoroughly and cautiously before making decisions about harmonizing tariffs.

The alternative option for solving the tariff problem, and the one that was written into the policy documents in Senegal before Inensus was formed, is to create a differentiated tariff structure. If this is implemented in practice, the private sector is thereby allowed to charge cost-reflective tariffs that account for the full cost of providing electricity in specific geographical areas. This has recently been done in Nigeria, in fact with the help of Inensus.

If the decision is to allow differentiated tariffs, policymakers should be upfront in communicating and explaining the logic behind the decision and why cost-reflective tariffs are considered important for delivering reliable and sustainable electricity access to wider and more remote geographical areas. The logic behind cost-reflective tariffs is typically that the electricity supply has a better chance of reliability and sustainable operation and economic performance in the long run if the revenue collected from the customers can cover all costs of operation and maintenance. According to this logic, such tariff setting is likely to be acceptable for the people in remote communities for several reasons: It brings electricity to areas where people are lacking access, it is the only available option for electricity supply in the areas apart from stand-alone systems, it gives much better quality lighting and other energy services than people have used before (kerosene, dry cell batteries, candles, etc.), and it costs less per day than the previous energy expenditure of many people. According to this reasoning, people will therefore appreciate the new services and also afford to use them.

A risk to this logic is that some policymakers or politicians will instead exploit the situation politically and promise lower tariffs. In general, it is common to use electricity tariffs politically, as in Senegal (see Chapter 2) where politicians promised low and harmonized tariffs to gain votes and keep social protest down in urban areas. At the same time, the economic management of the electricity sector was poor and required substantial support merely to rescue it from collapsing, instead of spending the same money on upgrading to avoid losses, expand more in rural areas, improve energy efficiency, and develop innovative solutions for affordable subscriptions for low-income groups.

There is no doubt that affordability is a great challenge for the work on universal access to electricity in developing countries and that a large number of people can only afford very low-cost subscriptions and tariffs, but this problem has to be tackled in new ways. For instance, as this book has shown, an emphasis on energy services and several levels of subscriptions can offer almost universal access in a rural village, especially to basic, important services such as light, phone charging, TV, radio, and fans, despite a higher price per unit of electricity than from the national grid. A current hypothetical argument, which is likely to be relevant to a long-term innovation perspective, is that if mini-grid developers achieve cost reductions that can make them able to match the national tariffs, this could solve many problems. It could avoid dissatisfaction in customers who are unhappy because their tariffs are higher than for their relatives in more central, grid-connected areas. It could also make mini-grid initiatives easier by removing the need to convince the people involved in the electricity access debate that mini-grid tariffs have to be higher than the national tariffs or, alternatively, not having to depend on cross-subsidization arrangements with differing degrees of suitability. However, we acknowledge the difficulties for mini-grid developers in matching national utility tariffs in the short-term, not least because the latter are subsidized in both visible and subtle, invisible ways. We also know that such key actors are working on this challenge.

The tariff issue has deep and entangled roots

The example from Senegal shows how difficult it can be for an individual company such as Inensus to impact national-level regulations despite committed efforts from the company. This is an example of the strong social structures in terms of path dependency in established technological systems (or socio-technical regimes) and the vested interests involved. Such inertia is a typical hindrance for radical change (socio-technical transitions), such as the emerging decentralized solutions for electricity supply (which can be seen as a socio-technical niche) within a regime mainly based on a centralized system. The issue in this case was not to get acceptance for the idea about solar mini-grids per se, but to get acceptance for a small-scale activity by a private sector company that could not use the existing tariff structure.

A paradox in the debate on harmonization of tariffs is that small stand-alone solar PV systems are not included, even though the electricity they deliver is more expensive per kWh than the tariff for grid electricity. However, this seems to be more accepted, while simultaneously also less visible and less comparable. The expectations of what mini-grids can deliver are higher than for stand-alone systems, because the mini-grids are more similar to the national grid.

For the main grid, the tariffs are usually not reflecting the real costs, and there is also cross-subsidization, sometimes invisible, from the areas where more electricity is sold to the areas with lower demand. Moreover, economic performance of national utilities is supported in various ways, and they are likely to be rescued if they perform poorly, as with the Senegalese utility. In addition, the energy authorities are often reluctant to extend the main grids to areas where they expect the demand for electricity to be low and increase only slowly. For private-sector-led mini-grids, since they are isolated projects, there is normally not such a flow of funds (visible or invisible) from central to remote areas. In such mini-grids, there is also more emphasis on connecting many people in each community and including the poorest than what is typical for main grid extension, even with recent initiatives to connect more people to existing grids. Therefore, the expectation of economic self-sufficiency for off-grid electricity provision is in a way unrealistic and unfair compared with grid extension.

Some of the barriers that at first sight look like regulatory challenges and technical discussions about tariff setting appear to be deeply rooted political problems. The debate about tariffs reflects a mix of different ways of thinking: equity-related thinking about affordable electricity prices, considerations about what is politically acceptable and popular among the voters, and deeper power relations and economic interests, placed at the national as well as the global levels. These have strong links to global financing institutions' historical and still ongoing pressure and interference in the political economy at the national level, with the World Bank as the most visible actor in our data. A part of the historical background is the promotion of foreign direct investment (FDI) during the last three decades as key to creating economic growth, as well as the liberalization of the energy sectors. This policy did attract FDI in African countries, but it has not been effective for social and economic development (Newell and Phillips 2016). Attracting FDI to a sector has tended to be seen as a measure of success in itself, regardless of its effects, according to Wethal (2017). In the Senegalese energy sector, FDI is present in terms of concessions for rural electrification and a few "independent power producers." The World Bank has pushed for concessions and a certain tariff model, while the government seems to partly resist this by keeping the previous model of state-led electrification, operating in parallel with the FDI model. It seems like the government wants the benefits of both and struggles with the disadvantages of both.

Inensus' initiative, supported by GIZ, was different from both of these, very small, and perhaps not interesting for the government, especially not when large donors started to offer funding for hundreds of mini-grids to be implemented in ways that the government was familiar with – tenders for equipment supply and installation, and tenders for operation and maintenance. This model, however, has been tried in Senegal earlier, and has met large problems of long-term operational and economic sustainability, as explained in Chapter 2. It seems like economic and political structures are rigged for large transnational international companies and donor interests, not for small-scale initiatives and companies as Inensus.

Reflections on the framework of analysis

The analysis of Inensus' long journey in Senegal has shown that all the six dimensions of the analytical framework are important for the understanding of the conditions for success with small-scale solar and hybrid mini-grids. The framework is likely to be useful also for other renewable energy provision at the village scale.

The first dimension, National and Global Energy System Context, provided the initial motivation for Inensus to create their socio-technical design and implement it in Senegal. There were signs that such a model would be welcomed and fit in well. This dimension was also where the fatal problems emerged. A transition from fossil fuels to new renewables like solar PV is not only a challenge of replacing technologies or getting acceptance for decentralized electricity provision, but also getting acceptance for new kinds of actors in electricity supply and developing new ways of thinking about the economic design and economic sustainability. This may include smart and reliable subsidies as a remedy for achieving affordability for people living in poverty.

The second dimension, Local Context, was a necessary background for analyzing the suitability of Inensus' mini-grid model in this context. Like many other actors developing decentralized electricity services, they had a genuine wish to develop solutions that were well adapted to the local context, including peoples' practical needs, economic situation, ability to pay, and opportunities for income generation. After implementing the socio-technical design (the third dimension), a dynamic interaction started between these two dimensions. The outcomes in terms of dimensions 4 and 5, Functionality (operational and economic sustainability) and Electricity Services, emerged through this interaction between energy system and context. Intensive learning processes took place for the project implementers.

The case illustrates how project implementers often learn the hard way through intense struggles and unpredictable outcomes. Such mechanisms should be acknowledged much more by other actors, and learning processes should be supported and facilitated, as suggested by Ockwell and Byrne (2017) in their important book on "socio-technical innovation system building".

Certainly, as stated in the literature, technological change is a social process, messy, unpredictable, and contingent on social contexts at different scales, but can be influenced by committed actors like the ones we have followed in this book. Implementation of small-scale, renewable mini-grids requires system innovation at multiple levels and depends on the long-term efforts of different kinds of actors.

Not only the interactions between the mini-grids and the larger energy system context at different scales, but also the micro-level dynamics between the mini-grid system and local context are important for the replicability of mini-grids. According to this research, adaptation of the socio-technical design to the local, socio-cultural, and material context is crucial to the embedding of the electricity system in this context. This is important to contemplate not only during the planning process but also over time after implementation, based on feedback from the people and lessons learned. In the Enersa case, for instance, the project implementers started to offer advice on energy-efficient refrigerators and other devices to make it possible for people to succeed with ice businesses. However, several requests from the population, including metered electricity and lower tariffs, were difficult or impossible for Enersa to accommodate within the existing system. Even if project implementers are very committed and have altruistic values in addition to the business goals, it might be difficult for them to meet all the needs of the people.

It may be asked whether electricity systems should differ in distinct geographical areas in order to make them sufficiently well adapted to the local context. This might appear to be in opposition to the need for upscaling and efficiency. However, similar socio-cultural contexts may exist over large geographical areas. The Enersa mini-grid model, for instance, would probably be suitable in several parts of Senegal. An energy system that is adapted to a specific un-electrified rural area with a high poverty level would also be likely to fit relatively well in another rural area with similar characteristics. However, some adjustments would probably be needed, because internal differences between villages, districts, and regions are likely. In the Enersa case and previously studied cases, there have been differences even from village to village. This means that the models used and the implementation strategies should be adjustable, via some inbuilt flexibility. In addition, regional differences in wealth levels, wealth distribution, livelihoods, settlement patterns, family constellations, and social norms are often large. Therefore, the energy models used in different contexts sometimes have to be significantly different (see Ulsrud 2015).

A sub-niche of decentralized electricity provision

As mentioned in the introduction, the Enersa case can be viewed as a project that contributes to system innovation in socio-technical niches, a so-called sustainability experiment. Typically, such experiments contribute to learning processes on what kinds of social and technological configurations that could work on the ground, which institutional changes they require, and which challenges and barriers need to be addressed. Such processes are important for strengthening emerging (energy) systems in niches and may eventually, in the long run, contribute to transitions, as explained in Chapter 1.

In this research, our main way of using theories was to build on previous insights in order to get ideas about which kinds of aspects or dimensions of the case could be important to investigate in order to understand it as a whole. We integrated the theoretical concepts in an open framework that could also capture other kinds of factors than those mentioned by theories. Our case is placed a step below the niche level, at the level of the individual sustainability experiment. Our framework of analysis is therefore placed at a lower level of abstraction than theories on socio-technical innovation. It focuses on the specific features of the projects, how they work and for whom, rather than on the degree of niche-building outcomes of the project. However, the project's outcomes in terms of niche building or system innovation are also important, and perhaps a seventh dimension of the framework should have been included. Inensus' lessons learned in Senegal have contributed to system innovation in several countries in Sub-Saharan Africa and elsewhere.

Solar and hybrid mini-grids can be viewed as a separate and very different niche than other off-grid renewables, such as solar home systems. Provision of electricity through small-scale, private-sector-led mini-grids has several similarities with provision through the main grid but also large differences. Especially the similarities in terms of connecting customers with grid lines and selling electricity appear to make this niche more affected by politics and more dependent on suitable regulations and cooperation with the government. In other words, the mini-grid model meets some challenges related to the political economy of states that sales or leasing of products like solar home systems and solar lanterns do not. The visions, networks, and learning processes are also very different in this niche than for individual solar systems. *It is probably reasonable to say, based on the discussion in this chapter, that the mini-grid niche challenges the established structures more directly than do household systems, at the same time as it is more dependent on relating to these structures. An implication of this is that the success of this niche (or sub-niche) depends on convincing governments and actors with influence on governments about the usefulness and potential of this kind of electricity provision.*

Recommendations

The focus of this research has been the activities of engaged actors attempting to make technologies useful in practice on the ground, striving towards important goals like economic sustainability and reliable, flexible electricity services. Moreover, we have investigated how the outcomes of these efforts are affected by the other involved actors and a range of contextual factors. The book clearly shows that the visions and efforts of the main driver and entrepreneur are crucial. However, no matter how committed they are, these efforts are not enough. Policymakers, donors, and investors are also important actors. Their task is both challenging and interesting because it entails changing structures, getting out of path dependency, and carrying out institutional innovation. We suggest recommendations for these different kinds of "facilitating" actors, since decentralized electricity provision by the private sector is a complementary and vital contribution toward the ambitious goal of universal electricity access.

Policymakers and regulators have a tremendous opportunity and responsibility for developing policies and regulations for private-sector-led mini-grids. Clear and tailored policies, laws, and regulations are necessary for private sector investment in mini-grid systems. Continued uncertainty in such conditional factors might lead to great economic losses for project implementers or effectively prevent them from contributing. Policymakers will have to address many challenges, most notably tariff setting, which is discussed in a dedicated section above, and the corresponding licensing and permitting. The goal for such policies and regulations must be to create conditions under which private investors see their investments and interests sufficiently protected. A particular issue is what happens when the national grid is extended to an area served by a mini-grid. Regulations must ensure that even in the event of national grid extension to a given site, the investors do not suffer economic losses.

Other important contributions from policymakers and regulators would be to work for simpler, faster, and less costly procedures for the companies that apply for licenses for electricity distribution. Innovative governmental institutions and support systems could be developed at several levels of governance, such as the district, county, regional, and national levels. Moreover, plans should be made for how to allocate areas between different actors, including the government. Another very important task would be to develop new ways of linking electricity access with different kinds of development initiatives, such as rural livelihood programs, roads, and transport, cash transfer programs, health, water supply, and sanitation. An open question is which kinds of organizations within and outside the governments should take the lead on coordinating and linking such different initiatives. New ideas for how the actors can develop their capacities accordingly are also needed.

Donors/international organizations have a potential to contribute in new ways. Previously, those donors/organizations that have supported project or business development on decentralized electricity provision have focused to a large extent on technical and economic assistance. No doubt, donor support has enabled a wide range of projects in diverse physical and socio-cultural geographies, testing and implementing different models. However, other kinds of contributions would also be useful, especially to the issue of working out novel policies and regulations. A promising example from Kenya is that GIZ is contributing to the work of the Energy Regulatory Commission to develop regulations for mini-grids developed by the private sector and other non-governmental actors. Other potential contributions by donors would include bringing together actors to share lessons; facilitating exchange of knowledge between countries; and offering information and education to government officials, policymakers, and others, while striving to learn from success and failure of capacity building in the energy sector in the past.

A useful kind of support for the energy providers that also benefits the people served by them is practical and economic assistance to facilitate income generation in the village.

Investors are encouraged to continue and expand their support to private sector initiatives. This is crucial for entrepreneurship, innovation, and experimentation as well as creating and harnessing future business opportunities. Such investments might lead to social and economic development in local communities, especially if they are performed in combination with other tangible efforts toward poverty alleviation and holistic rural development.

One of the most important tasks of the *project implementers*, as has been discussed much more in Chapter 8, would be to work on cost reductions, including the overhead costs, and the challenge of balancing this with flexibility for the users, facilitation of income generation, and other factors presented in Chapter 8. These key actors are also central in encouraging, inspiring, training, and working closely with all the other actors mentioned above in order to create trust and foster creativity and mutual exchange of ideas, and this is part of Inensus' ongoing work. Inensus has also become a facilitator for other small companies, such as in Nigeria.

For the *citizens of the villages*, our recommendation is to continue to push the project implementers to understand the local conditions, needs, and constraints and to come up with new ideas for how the electricity supply should work and how it can be linked to income generation and jobs, health, water supply, and education.

References

Bloomberg New Energy Finance. (2016). Off-Grid Solar Market Trends Report 2016. Market Analysis.

Newell, P. & Phillips, J. (2016). Neoliberal energy transitions in the South: Kenyan experiences. *Geoforum*, 74, 39–48.

Ockwell, D. & Byrne, R. (2017). *Sustainable energy for all: Innovation, technology and pro poor green transformations*. New York: Routledge.

Raven, R. P. J. M. (2005). *Strategic niche management for biomass: A comparative study on the experimental introduction of bioenergy technologies in the Netherlands and Denmark*. Eindhoven Centre for Innovation Studies, VDM Verlag. 340 pp.

Ulsrud, K. (2015). Village-level solar power in practice: Transfer of socio-technical innovations between India and Kenya. *PhD dissertation*. Department of Sociology and Human Geography, Faculty of Social Sciences, University of Oslo.

Wethal, U. (2017). When China builds Africa: Linking construction projects and economic development in Mozambique. *PhD dissertation*. Department of Sociology and Human Geography Faculty of Social Sciences, University of Oslo.

Acknowledgements

We are very grateful for the research funding from the Research Council of Norway, through grant number 217137. Our sincere thanks further go to the people who have shared their experiences, views, and information with us for this research. The willingness of key actors to be open and share the details of their work, including the parts that went differently from what they had hoped, has been very important for the insights achieved. Special thanks to Jakob Schmidt-Reindahl for patiently following up on emerging needs for additional information. We would also like to thank our colleagues in the Solar xChange team who have contributed to this work, including Henry Gichungi, Jonas Sandgren, Karen O'Brien, Harald Rohracher, Tanja Winther, Kristen Wanyama, and Cecilie Fardal Nilsen. Moreover, Hege Sørreime has provided useful comments on the manuscript and generously shared her office during the finalization phase.

Index

Printed and bound by CPI Group (UK) Ltd, Croydon, CR0 4YY

28/10/2024

01780061-0001